Contributions to Law, Philosophy and Ecology

Contributions to Law, Philosophy and Ecology: Exploring Re-Embodiments is a preliminary contribution to the establishment of re-embodiments as a theoretical strand within legal and ecological theory, and philosophy. Re-embodiments are all those contemporary practices and processes that exceed the epistemic horizon of modernity. As such, they offer a plurality of alternative modes of theory and practice that seek to counteract the ecocidal tendencies of the Anthropocene. The collection comprises eleven contributions approaching re-embodiments from a multiplicity of fields, including legal theory, eco-philosophy, eco-feminism and anthropology. The contributions are organized into three parts: 'Beyond Modernity', 'The Sacred Dimension' and 'The Legal Dimension'. The collection is opened by a comprehensive introduction that situates re-embodiments in theoretical context. Whilst closely bound with embodiment and new materialist theory, this book contributes a unique voice that echoes diverse political processes contemporaneous to our times. Written in an elegant and accessible language, the book will appeal to undergraduates, postgraduates and established scholars alike seeking to understand and take re-embodiments further, both politically and theoretically.

Ruth Thomas-Pellicer is an independent scholar based in Catalonia, Spain.

Vito De Lucia is a Research Fellow in K. G. Jebsen Centre for Law of the Sea, Faculty of Law, UiT – Arctic University of Norway.

Sian Sullivan is Professor of Environment and Culture at Bath Spa University, UK.

Law, Justice and Ecology
Series editor: Anna Grear
Cardiff Law School, Cardiff University, UK

In an age of climate change, scarcity of resources, and the deployment of new technologies that put into question the very idea of the 'natural', this book series offers a cross-disciplinary, novel engagement with the connections between law and ecology. The fundamental challenge taken up by the series concerns the pressing need to interrogate and to re-imagine prevailing conceptions of legal responsibility, legal community and legal subjectivity, by embracing the wider recognition that human existence is materially embedded in living systems and shared with multiple networks of non-humans.

Encouraging cross-disciplinary engagement and reflection upon relevant empirical, policy and theoretical issues, the series pursues a thoroughgoing, radical and timely exploration of the multiple relationships between law, justice and ecology.

Titles in the series:

Law and Ecology
New environmental foundations
Andreas Philippopoulos-Mihalopoulos

Law and the Question of the Animal
Edited by Yoriko Otomo and Edward Mussawir

Wild Law – In Practice
Edited by Michelle Maloney and Peter Burdon

Earth Jurisprudence
Peter Burdon

Contributions to Law, Philosophy and Ecology

Exploring re-embodiments

Edited by Ruth Thomas-Pellicer, Vito De Lucia and Sian Sullivan

Routledge
Taylor & Francis Group

LONDON AND NEW YORK

First published 2016
by Routledge

2 Park Square, Milton Park, Abingdon, Oxfordshire OX14 4RN
711 Third Avenue, New York, NY 10017
a GlassHouse book

Routledge is an imprint of the Taylor & Francis Group, an informa business

First issued in paperback 2017

British Library Cataloguing in Publication Data

A catalogue record for this book is available from the British Library

Library of Congress Cataloging-in-Publication Data
Names: Thomas-Pellicer, Ruth, 1974- editor. | De Lucia, Vito, editor. |
Sullivan, Sian, editor.
Title: Law, philosophy and ecology : exploring re-embodiments / edited by
Ruth Thomas-Pellicer, Vito De Lucia and Sian Sullivan.
Description: Abingdon, Oxon ; New York, NY : Routledge, 2016. | Series: Law,
justice and ecology | Includes bibliographical references and index.
Identifiers: LCCN 2015043504| ISBN 9781138852877 (hbk) | ISBN 9781315723235
(ebk)
Subjects: LCSH: Philosophical anthropology. | Philosophy, Modern. | Ecology.
| Law--Philosophy.
Classification: LCC BD450 .L355 2016 | DDC 128--dc23
LC record available at http://lccn.loc.gov/2015043504

ISBN: 978-1-138-85287-7 (hbk)
ISBN: 978-0-8153-8520-2 (pbk)

Typeset in Baskerville by
Fish Books Ltd.

Contents

List of illustrations

Notes on contributors

Patrick Curry is an independent scholar living in London. He is the author of several books including *Ecological Ethics* (revised edition 2011) and is currently writing a book about enchantment. He has held lectureships at the universities of Bath Spa and Kent. Most of his papers are available on his website (www.patrickcurry.co.uk).

Vito De Lucia is a Research Fellow at the K. G. Jebsen Centre for Law of the Sea, Faculty of Law, UiT – Arctic University of Norway. His research interests include international environmental law, climate justice, critical ecological approaches to law, and ecological legal theory/philosophy.

Sue Farran is a Professor of Laws at Northumbria University, Newcastle, England. Her research focuses on the intersection of law and people and challenges to dominant discourses in different contexts, drawing in particular on post-colonial development in small Pacific island states and the relationship between different players in shaping and developing legal responses to contemporary issues.

Arran Gare is Reader in Philosophy at Swinburne University, Melbourne, Australia. He is the author of a number of books, including *Postmodernism and the Environmental Crisis* (London: Routledge, 1995) and *Nihilism Inc.: Environmental Destruction and the Metaphysics of Sustainability* (Sydney: Eco-Logical Press, 1996). In 2005 he founded the on-line journal *Cosmos and History*.

Dr **Adrian Harris** is a psychotherapist and embodiment theorist. His PhD explored embodied knowing in Eco-Paganism and his MSc researched psychotherapy in nature. Adrian has published on animism, embodied knowing and the power of place. He edits Embodiment Resources (www.Embodiment.org) and blogs on the relationship between bodymind and place: www.bodymindplace.org.

For over forty years, **Madronna Holden** taught philosophy, anthropology, folklore, ethnic studies, and women's studies at universities in the US and abroad, and is internationally published in these fields. Some of her lectures are available online on her website, Our Earth/Ourselves (https://holdenma.wordpress.com/), which has had visitors from 180 countries.

Gabrielle O'Shannessy is an Honours Graduate in Law from Southern Cross University, Australia. Her interests are in redefining our notion of self to encompass holistic and ecological ontologies and how such ontologies may be incorporated within legal systems. She is also a musician and animal lover and is currently working as a lawyer in Northeastern New South Wales.

Dr **Alessandro Pelizzon** is a Lecturer in the School of Law and Justice at Southern Cross University. Alessandro's areas of expertise are legal anthropology, comparative law, legal theory, indigenous sovereignty, and ecological jurisprudence.

Margherita Pieraccini (MA Aberdeen, Mphil Cambridge and PhD Newcastle) has been a lecturer in law at the University of Bristol Law School since 2011. Margherita's research is interdisciplinary, bridging environmental law with social anthropology. Her research focuses on nature conservation, the commons, and legal pluralism in European countries.

Sian Sullivan is Professor of Environment and Culture at Bath Spa University, UK. Her research explores cultural landscapes, the financialisation of nature, the politics of biodiversity conservation, and embodied relationship with nature-beyond-the-human. Much of her work can be found online (http://siansullivan.net).

Dr **Ruth Thomas-Pellicer** is an independent scholar intending to bring about a post-ecocidal turn in the fields of philosophy and science. Her claim is that Western metaphysics has proved ecocidal through and through and must therefore be strategically overcome. Thomas-Pellicer's approach is utterly cross disciplinary.

Ali Young completed her doctorate in leadership in 2014 at the Centre for Leadership Studies in Exeter, UK. Her interests include embodiment, spirituality, ecofeminism, aesthetics, and constructive postmodernism. She also teaches Movement Medicine, offering workshops in embodied spiritual practice, employing a combination of dance and shamanic teachings, with the intention of promoting greater commitment to the well-being of all life.

Acknowledgements

We are grateful to the British Sociological Association Theory Study Group for sponsoring a workshop on the theme of re-embodiments, which became the starting basis of the present volume. Similarly, we are thankful to Anna Grear, director of this book series, for her firmness in suggesting to give form to the present collection.

Series editor's introduction

There is a sense in which nothing could be more urgent for addressing the intersection between law, justice and ecology than a sustained engagement with what we could think of as the politics of disembodiment/embodiment. It has long been recognised that trajectories of disembodiment both inaugurated and characterise the familiar, patterned and highly predictable forms of injustice implicated in what Merchant famously named 'the death of nature'.[1] Longstanding socio-ecological practices of predation, exclusion and appropriation – the very practices extensively underlying the emergence of the Anthropocene problematic – centrally assume – as multiple critiques demonstrate – a disembodied human subject for which 'everything else' (including the subject's own body) is mere 'environment'.

It is in response to this central problematic that the present collection can be located. The collection presents one lively thread in an on-going exploration of – and search for – alternative ways of thinking and imagining 'the world' to shape eco-humane 'future histories' worth living.

This collection, in multiple voices and from various positions, animated by differing disciplinary and theoretical energies, converges to explore 're-embodiments'. Re-embodiments are presented in this collection as 'a set of *counter*-trajectories that explore alternative ways of imagining the real, and of articulating it in philosophical, legal and ecological terms'.[2] Alternative, that is, to the disembodying dynamics at the heart of modernity.

Turning away from disembodying impulses and trajectories is surely a central task for scholarship in an age of multiple eco-crises – as is grappling with the sheer liveliness of the materio-semiotic complexity revealed by new scientific insights, the intensifying inter-permeations between the global and the local and the dazzling pace of boundary-collapsing developments in technoscience. By framing 're-embodiments' as emergent relations within a wider, dynamic ontological field, this collection can be seen as a distinctive thread in a much broader contemporary concern with processes and relations – with interpermeation, entanglement and radical interdependencies. This collection, ultimately, explores the foundations for a vision reaching beyond the trajectories of contemporary crises, beyond

ecocide, towards a new imagining of the philosophical and political implications of lively materiality itself.

Anna Grear

Notes

1   C. Merchant, *The Death of Nature: Women, Ecology and the Scientific Revolution* (New York: Harper Collins 1989).
2   Introduction.

Introduction

Exploring re-embodiments

Ruth Thomas-Pellicer and Vito De Lucia

Introduction: 're-imagining our contemporaneity'

This collection began as a workshop convened by Ruth Thomas-Pellicer under the aegis of the British Sociological Association (BSA) (Theory Study Group) in 2010. The title of the workshop was 'Re-imagining our Sociological Contemporaneity: What is the Age of Re-Embodiments?',[1] and the central exercise that the workshop contributors were asked to engage with was one of *re-imagination*. 'Re-imagining our contemporaneity' was, to be sure, the core framing of the call for papers. This re-imagination was oriented towards the conceptual referent of 're-embodiments'. In this respect, the original call for papers spoke tentatively of an *Age* of Re-Embodiments as the horizon for the imaginative engagements the call sought to invite. Not surprisingly, some papers in the present collection still explicitly retain the phrase 'Age of Re-Embodiments'.[2]

Re-embodiments, in plural, intends to capture the nuanced, complex and situated multiplicity of re-embodiments, a multiplicity of perspectives whose exploratory exuberance – we think in hindsight – cannot be easily contained within a coherent, unitary, frame such as an 'age', in either the temporal or conceptual sense of the latter. Within this more-nuanced context, even the quest for an 'age', for a new grand narrative that may counter and reverse modernity's ontological and epistemological trajectory of disembodiments, can be understood as *one* thread in what, ultimately, is a collective, as yet unstable, weaving effort.

It must be nonetheless observed that the notion of 'age' is somehow refracted back to the collection through the forcefully emerging frame of the Anthropocene: what is being proposed as a veritable *geological age* whose hallmark is the rise of humanity as a geological force, and to which we will return later in this introduction. Even this geological age, however, is arguably the result of a plurality of trajectories operating at different spatio-temporal scales.[3] Re-embodiments then may be usefully understood as a set of *counter*-trajectories that explore alternative ways of imagining the real and of articulating it in philosophical, legal and ecological terms.[4] In this respect, there remains a productive tension traversing the edited collection at hand between a poststructuralist problematizing stance, and a post-postmodern desire for carving theoretical space for a new project that escapes both modernity and its postmodern intensification.[5]

At the semantic level, the framework of re-embodiments marks a specific difference with respect to the *dis*embodying trajectory of modernity.[6] At the conceptual level, re-embodiments exceed the material, though inevitably crucial, linkage with the body, by emphasizing, ultimately, processes and relations. Embodiments, in fact, are understood as hubs or *nexi* of relations, rather than delineated ontological entities, as we shall see. At the grammatical level, through its plural articulation, the framework of re-embodiments marks a difference with respect to the ideological uniformity and homogenization modernity imposes to any form of complexity and plurality.[7]

But before further discussing each of these distinctions, we must ask: why a new imagination? We will argue that a new imagination is required in order to respond – i.e., to act responsibly and responsively – to the ongoing ecological crises that are increasingly, and aptly, portrayed as ecocides. Indeed, an increasing number of reports on the dire state of the planet in relation to a number of growing and overlapping ecological crises – or, rather, ecocides – are readily available.[8] By ecocide we mean the large-scale, deliberate destruction of the natural environment, of ecosystems and, more broadly, of the planetary *oikos*.[9] The prevailing way to respond to the challenges imposed by the present situation hinges usually on technology[10] and markets.[11] Yet, we rather argue that what is necessary – and this resonates with an increasing amount of scholarship across multiple disciplines – is, in the words of Anna Grear, a 'radical, collective worldview-shift'.[12] The possibility of a new imagination hinges then on the desarticulation of a set of key conceptual referents through which Western modernity apprehends and constructs the world. This is due to the fact that imagination, being always socially and culturally situated, *represents* and *imitates*. In this respect, in order to imagine something novel, we must first dislodge the epistemological (and ontological) organizers of modernity.

The epistemic imaginary of modernity and the 'resolve' of the anthropocene

This section will read Western/ized culture, particularly in its modern articulation, as centrally structured around a particularly powerful organizing framework, a particular epistemic imaginary, which generates practices for knowing nature that, as interpreted through a Foucaudian lens, invariably operate as practices of power. Our claim, in this respect, is that the 'resolve' of the Anthropocene is the political enactment of the epistemic imaginary of modernity, which is, no doubt, inextricably interpenetrated with the workings of capitalism.[13] This epistemic imaginary is, in turn, organized around the central concept of *solvere* (introduced below). However, as we shall see below, *solvere* and resolve belong to the same semantic field. In this light, we can state that the Anthropocene is the performative dimension of the modern epistemology. The Anthropocene, that is, facilitates the continuous transformation of an epistemic imaginary of mastery (*solvere*) into a political resolve to dominate.[14] Accordingly, in the next two sections we will discuss, first, the epistemic imaginary organized around the frame of *solvere* and, subsequently, the Anthropocene.

Solvere as epistemic imaginary of modernity

Following Heidegger, we will explore *solvere* not through analysis, but rather through etymology. Analysis aims at 'tightening up' or 'narrowing' the meaning of a term (what Heidegger calls 'stunting the word').[15] Etymology, on the other hand, aims at '*opening up*' the word, in order to reveal the richness of its semantic field.[16]

Solvere is a Latin verb whose semantic field indicates broadly 'separation'. In its English translation it can mean 'loosen'; 'dissolve'; 'untie'; 'release'; 'detach'; 'depart'; 'unlock'; 'scatter'; 'dismiss'; 'accomplish'; 'fulfil'; 'explain'; and 'remove'.[17] This rich semantic field has propelled an entire philosophical framework – namely, analytical philosophy – which reduces thinking to the 'dissolving/resolving or eliminating/denying', through analysis, of 'problems'.[18]

It must be further highlighted that *solvere* operates at an additional semantic level that is crucial for the stage of our storyline and that is perhaps closer to the common English usage of words such as 'solve' and 'resolve'. In fact, the etymological richness of *solvere* gives rise to words such as 'solve', 'solution', 'resolve', 'resolute', 'resolution', 'absolute' and 'absolve'. All these words, it must be noted, refer not only to the fundamental semantic character of separation, but also, importantly, to the crucial template of the modern subject. For the modern subject – to gain its modern edge – must be resolute, rational, autonomous and capable – in a material and moral sense alike – of both sovereign decision and the taking of control of the world through its will.

The implications of the horizon of sense implicated by the organizing framework of *solvere* are most immediately evident through the articulation of the particular epistemology of modernity. To be sure, this framing helped develop a new epistemological imaginary, which is perfectly captured by the new methodology elaborated by Galileo. Galileo's manner of proceeding through resolution–composition entailed a methodological reductionism by which science could decompose, reduce – *resolve* as it were – reality in its smallest components, to subsequently re-compose it in a logical fashion.[19] A veritable disembodying operation continuously re-enacted in the operations of science, as epitomized in the practices and aims of contemporary – which we may aptly qualify as 'anthropocenic' – [20] biological, molecular and nano-technologies.

Nevertheless, this epistemology required the construction of a knowing subject. In this respect the metaphor of *solvere* has recourse to an Archimedean point that becomes successively variously articulated: as 'God'; as the political theological concept of the sovereign (Hobbes, Schmitt, Agamben); as the subject owner (Locke); as the subject of human rights (Kant). Moreover, in its disenchanting secularized version, namely, in a scientific setting, the Archimedean point coincides with 'objective truth'. Objectification resolves and therefore destroys the precarious semantics intrinsic to metaphor.[21] It is in annihilating this quivering construction of truth that science gains its mastery.

Modernity sets up nature in an ontological sense such that it can be known epistemologically through the disaggregating practices of *solvere.* More-than-human nature thus appears as a number of relations that have been broken up and have been disintegrated. Yet, paradoxically, the non-human world is referred to under the uniform banner 'nature', and is rendered thusly as a homogeneous monolith, which is pitted against the equally monolithic realm of culture. It is only once this ontological and epistemological threshold is established as the overall operative framework, and nature is reduced to an assemblage of dead parts without agency or affectivity, that the re-solution of nature into its constituent parts can be enacted.

Thus we argue that the epistemic imaginary of modernity – as the current historical and theoretical articulation of the Western metaphysical trajectory – is largely (and hegemonically) organized around the frame and conceptual structure of *solvere.* Indeed, *solvere,* as a central epistemological referent, establishes a particular horizon of sense that is fundamental to the thinking and operations of two of the primary mechanisms of modernity: science and law.[22] It must be however noted that s*olvere* has two interpretive registers. On the one hand, it provides the broad epistemological framework within which reason is constructed, understood and operationalized. On the other, *solvere* is more immediately directed to a political project. In this, the etymological richness of the concept of *solvere* is condensed and sharpened in what can be described as the *resolve* of modernity to enact its epistemological identity, which, ultimately, translates into what we have called the resolve of the Anthropocene. This latter period enables the consummation of the continuously re-enacted transformation of epistemology – i.e. mastery – into practices of power – i.e. domination. But what is exactly meant by the Anthropocene?

The anthropocene as human intervention in geological time

The growing influence of humanity on the natural world was already described as a 'telluric force' in 1873 by Italian geologist Antonio Stoppani, who spoke of an 'anthropozoic era'.[23] More recently, Paul Crutzen, in a now-famous essay published in *Nature* and called 'The Geology of Mankind', the term 'Anthropocene' is called on to denote the depth, pervasiveness and permanence of humans as a distinctive geological shaping force.[24] As such, the Anthropocene – whose distinctiveness is that it is human dominated – has been proposed before the International Geological Commission to be adopted as a new formal geological epoch – the current epoch being the Holocene.[25] However, it has already become a significant conceptual framework whose usage has 'rapidly escalated',[26] and that has given rise to at least three dedicated academic journals.[27]

The Anthropocene informs an increasing number of radical scholarly reflections and ethico-political projects. However, these are bifurcated. On the one hand, the Anthropocene suggests that the distinction between natural and artificial no longer has any meaning; hence, we should impose a comprehensive system of management and control on the entire planet, in order to ensure its conservation, and even its enhancement.[28] Indeed, as a 'new' ecologist, Botkin suggests, "[n]ature in the twenty-first century will be a nature that we make; the question is the degree to which this molding will be intentional or unintentional, desirable or undesirable".[29] On the other hand, critical scholars argue that the Anthropocene requires, not an intensification of human interventionism, but a radical, paradigmatic shift in human theoretical endeavours.[30] Louis Kotzé argues in this respect that 'the arrival of the Anthropocene is possibly set to require a complete rethink' of the framework of human rights in relation to the environmental regulatory domain.[31] Similarly, a recent collection edited by Anna Grear and Evadne Grant titled *Thought, Law, Rights and Action in the Age of Environmental Crisis*, suggests the urgency of this task by underlining how '[i]n the climate-pressed Anthropocene epoch, nothing could be more urgent than fresh engagements with the fractious relationships between 'humanity', law and the living order'.[32]

Modernity and/as disembodiment

Within the conceptual and historical context of the Anthropocene, there is a further intensification of the central trajectory of modernity, namely disembodiment.[33] Disembodiment is in effect the crucial conceptual achievement of the Cartesian paradigm, which enacts a 'body excision',[34] and enables a progressive disembodiment – descriptive, as well as productive or performative – of which law is a crucial operator. Indeed, Cartesian disembodiment has provided the conceptual resources for the construction of a (legal) subject aligned with the organizing framework of *solvere*, to the extent that the Cartesian subject is thoroughly re-solved – that is, in line with the semantic field of *solvere*, dis-integrated and separated –

from its affective, social and ecological *situation,* and reduced to abstract thinking, 'naked existence' and 'bloodless reality'.[35]

In the political philosophy of both Hobbes and Locke – crucial well-springs of fundamental political and legal categories of modernity – the organizing framework of *solvere* is utterly at work through what Esposito calls the (biopolitical) 'paradigm of immunization'.[36] Hobbes' sovereign *dispositif* is the political form of the 'desocialization' of individuals, of the re-solution and dis-solution of the horizontal relational bond that keeps them in common.[37] 'Life', writes Esposito, is both 'privatized and deprived of that relation that exposes it to its communal mark';[38] every horizontal relationship that remains external to the 'vertical line that binds everyone to the sovereign command is cut at the root'.[39] Locke similarly constructs its proprietary *dispositif* in such a way that it operationalizes the frame of *solvere*. The re-solution and dis-solution of the communal bond is here activated through the proprietary frame. Property severs bodies and things from their field of relations and re-roots them exclusively within the sphere of *dominium* of the subject while simultaneously excising the communal bond with other subjects, and transforming subjects' interactions into what Pashukanis would describe as the 'formalized legal exchanges between commodity owners'.[40] This proprietary enactment of *solvere* in relation to bodies and things of the world still provides the epistemic imaginary operative in law,[41] and it is exactly what some legal scholarship has begun problematizing through a series of attempts at thinking law ecologically.[42]

The construction of the legal subject as a *persona* masking the concrete particularities of the individual humans hinged indeed on the transformation of the complex modalities of human interaction into a rarefied legal mode that presupposed the erasure of the embodied, experiential materiality of men and women.[43] As Loick observes, the concept of persona, which originally means mask, 'hides the uneven features of individual faces and places upon all an equal, even shape'.[44] From this masking, from this encryption of difference, 'a gap opens up between the existential wholeness of particular individuals as they exist concretely in the world (homo) and their juridical avatars (persona)'.[45] It is precisely this gap, which, through a slippage, renders operative the exclusionary template of legal subjectivity in a way that will be fully developed only with Descartes and then with Kant.[46]

The slippage consists in the fact that, if the concrete materiality goes out of the door through the device of the *persona,* it comes back, so to speak, through the window of its *maintaining* a link, however tenuous, with a particular version of humanity – rational, white, male... and here the second semantic register of *solvere* is decisively operative.[47] It is through this slippage that the exclusionary template is enacted: first by 'withholding the protection of the legal mask'[48] from recalcitrant beings incapable or unwilling to approximate the abstract template – women, savages, blacks, nature, etc.; and then intensified through the deployment of human rights in relation to corporate legal entities, that is, fictitious persons entirely devoid of the human substrate – and because of this lack, offering to be a perfect match for this masked legal subjectivity.[49]

This mechanism then, that is '[t]he institution of personality, [produces] a completely opaque space, in which one member of the community [is] inaccessible to another'.[50] Hence the re-solving and dis-solving of the social bond between humans in community – a further expression of how the epistemic horizon of *solvere* is enacted through such onto-juridical devices.

Errancies of modernity: thinking re-embodiments through *religare*

Nietzsche and Heidegger suggest that modernity is an indefatigable race marked by novelty and constant change. This incessant 'overcoming', this 'unstoppable movement' towards the increasingly rapid obsolescence of the new[51] enacts the operative metaphor of *solvere* in a declension that continuously solves and dissolves the present. Marx and Engels similarly alerted in the *Communist Manifesto* to the unparalleled restlessness exhibit-ed on the part of the bourgeoisie in their eagerness to develop new devices and the resulting urge to constantly expand markets.

In a framework characterized by its own proneness to unrelenting catching up with itself, the strategy of 'overcoming' to exit modernity loses its critical purchase. Unlike the claims of accelerationist positions that advance that capitalism should be expanded in order to generate radical social change,[52] Nietzsche and Heidegger both state that the decentring of the modern trajectory does not consist in further moving forward in a straight line. Rather, the destabilization of modernity is most usefully attempted through multidirectional incursions into the grounds and tracts that the former has banned: it is the areas that modernity has qualified as untruthful that need revisiting and reconsideration.

This plurality of strategies, this ensemble of counter-offensives, can be imagined, with Nietzsche, as the 'errancy' of Western metaphysics of which modernity is an intrinsic part. At the end of the metaphysical trajectory, Nietzsche suggests, 'thought does not return to the origin in order to appropriate it; all that it does is travel along the multiple paths of errancy, which is the only kind of wealth [...] that is ever given to us'.[53] Modernity's decentring thus hinges on the (re-)activation of knowledges and practices that have been silenced and repressed during its steady ascendancy.[54]

Re-embodiments therefore, we suggest, can be imagined as consisting of plural trajectories of errancy, such as the ones explored, (re-)activated and articulated by the contributions in this volume: ecological civilization (Arran Gare); animist practices (Ruth Thomas-Pellicer, Ali Young, Patrick

Curry and Adrian Harris); ecocentrism (Patrick Curry as well as Alessandro Pelizzon and Gabrielle O'Shannessy); ecofeminism (Madronna Holden and Ali Young); historically situated socialization processes (Sian Sullivan); collective and commons practices (Vito De Lucia and Sue Farran); indigenous practices (Madronna Holden); and social-ecological practices of co-production of law (Margherita Pieraccini). At the same time, however, we argue that these trajectories, even as they remain plural, can be folded within a common horizon of sense that, we propose, is well captured through the organizing frame of *religare*. Like *solvere*, *religare* operates as the conceptual hinge on/through which an epistemic imaginary is articulated. While *solvere*, however, dis-solves, separates, reduces, dis-integrates and disembodies, *religare* brings together, (re-)integrates, situates and, as we shall see, most importantly, re-embodies. To be sure, it is through *religare*, we propose, that we can think re-embodiments.

Thinking re-embodiments through *religare* offers two interlinked interpretive registers. One is linked most immediately with the body and its materiality, in a manner contiguous with new materialism's re-ascription of vitality and agency to bodies and things.[55] This is, no doubt, a crucial passage. The second interpretive register expands this first perspective by understanding the 'body' outside of the mind/matter binary of modernity. We shift then to a language of embodiments. We do that because, we advocate, ascribing agency to the materiality of things, understood as firmly, stably bound bodies, is not sufficient, not for our purposes. The embodiments underlying re-embodiments, and this, we suggest, marks a crucial difference, are clusters, hubs or *nexi* of relations whose provisional ontology – embodiment – acts most immediately in ethico-political terms – as either *dis-* or *re-*embodiment, though the complexities involved may elude a uniform reading. An embodiment, from this perspective, transversally integrates a multiplicity of relations of which it becomes a hub, a point of materialization, an interpretive key.

In the next sections, we will first try to explain in more details the concept of embodiment, its double interpretive register and its theoretical and ethico-political usefulness; and subsequently, we will articulate the framework of *re-*embodiments as a plural ethico-political project. After that, we will further explore the conceptual framework of *religare*.

From bodies to embodiments

In the Cartesian–Newtonian tradition, matter and 'corporeal substance'[56] have been put on a par. These were 'extended, uniform and inert'.[57] As a result, they were amenable to quantification and manipulation in line with Euclidian geometry and Newtonian physics.[58] Moreover, bodies are thusly assigned a precise ontological delineation that makes them bound entities isolated and isolable from the situation in which they are immersed and from the relations through which they are constituted. The focus on the affective presence, vitality and agency of bodies that feminist and new materialisms scholarship emphasize is in this respect a central element of re-embodiments. Indeed, bodies are sites of subjugation, occupying spaces of intersection where disciplines and power relations are inscribed and find material concretization. Bodies are also sites of resistance, through counter-hegemonic and non-hegemonic practices. We argue, however, that the key to re-embodiments is a passage from bodies to embodiments. This linguistic shift aims at displacing the binary that hides, inevitably like a shadow or a ghost, within any conceptual redeployment of the term body. A body, we fear, always indicates the absence of its counterpart, namely, the mind. Shifting emphasis from mind to body, from reason and ideas to materiality, we claim, is insufficient.

By contrast, embodiment offers flexibility and distance from the shadows and ghosts of modern metaphysics, as it indicates, as announced above, first and foremost a hub or *nexus* of relations. Importantly, embodiment also denotes any sort of provisional materiality – a human body, a non-human body such as a tree, a stone, a forest, a technological artefact, an automobile, a house, a city. Each embodiment comprises, in this respect, both material structures and symbolic resources; it is at once natural and cultural; it is ontologically present but resists ontological closures; it embodies particular ethico-political practices that may either facilitate and expand or resist the dis-embodying effects of the modern homogeneous resolution of relations. Yet embodiment is more.

Embodiment is amenable to neither ideal nor molecular reduction. As a result, the essentialist issue of its constitutive nature – whether ultimately ideal or material – is largely inconsequential to the perspective defended in the frame of this collection. Crucially, as it is relations that structure embodiments, no embodiment enjoys a vantage position vis-à-vis another embodiment. Lacking an Archimedean point from where to operate, no objectification, in the modern sense, can take place. Embodiments structure thus an interminable immanent mesh from which they singularly emerge as hubs or *nexi*. An embodiment thus is no longer necessarily an individual, 'in the modern sense of an undivided (and hence undividable) unit'.[59] We are therefore requested to pay attention to the relations that traverse embodiments, for these are largely constitutive of the latter. It is in these interactions where the local effects of the objectifying Western metaphysical project in all its assorted forms – thoroughly apparent in both modern science and law – are inscribed.[60] This perspective allows us to reveal the extent to which ideas, policies and acts unleash disembodying or re-embodying effects; at the same time, it displaces the question of where to locate the ontological closure, and aims at entirely refracting the threshold between nature and culture. The question then is no longer 'what is Enlightenment?', but, rather, 'what are the ecological effects of

Enlightenment?'; it is no longer 'what is that embodiment?', but, rather, 'what are the ecological effects of that embodiment?' Ecology, it must be emphasized, is here understood, with Guattari, in a transversal sense – that is, investing the natural, the social and the psychological at once.[61] In this light, we can state that re-embodiments are an ethico-political project.

Re-embodiments as a plural ethico-political project

Re-embodiments are practices – descriptively – and attempts – ethico-politically – at undoing, to the possible extent, ecocidal practices. As such, re-embodiments convey an affirmative attitude that is grounded in practice. They put forth a way of living – re-embodying and as a result re-embodied – which deflects at least some of the binary pairs endemic to the Western trajectory, chief among which are the subject/object partition and the idealist/materialist division. '"There is no being", [Nietzsche] wrote, "behind the doing, acting, becoming [...] the doing is everything"'.[62] Re-embodiments are indeed 'doing, acting, becoming'. *Qua* deeds or activities, re-embodiments emerge as truly a-Cartesian. While embodiments, as we have seen, start from the materiality of the body – including, importantly, its vulnerability – they do not offer particular ontological closures, but rather an *opening up* towards the mutually implicating relations that each provisional embodiment speaks of and enacts. In this respect, embodiment emerges, as noted above, as an ethico-political category and *re*-embodiments as a plural ethico-political project.

The plural ethico-political inflection of the concept of re-embodiments connects re-embodiments as practice and project to the philosophical tradition that Esposito calls 'living thought',[63] traversed by three central elements. First, the modern 'abrasion of the origin' is challenged through a genealogical reading of history.[64] Modernity took hold through the establishment of thresholds – anthropological, epistemological, onto-logical, institutional – in order to negate the continuity between 'man' and the animal, natural world, and to affirm the liberation from 'man's' dark, chaotic origin in 'nature' through a new rational, artificial beginning.[65] The philosophical tradition of living thought genealogically rejects the myth of the origin – an origin paradoxically invented in order to displace and forget it, as its biological materiality was perceived as a threat, a theme perhaps most forcefully expressed in the political philosophy of Hobbes. The entire modern tradition that responds to the practice of knowing, of *solvere*, enacts the abrasion of the origin through the marking of thresholds.

Descartes marks the threshold that separates the body from reason. Hobbes, in what will become a trope, marks a threshold between the state of nature and civilization: in the former, humans live in 'continual feare, and danger of violent death';[66] this fear, for Hobbes, can only be neutralized through a 'common power' that keeps all men 'in awe', that is, the sovereign *dispositif*.[67] Each threshold, however, aims at constructing and protecting human life, through reason, from the excesses and dark forces of life. Yet in so doing, the origin, which can never be neutralized, only returns cyclically in a spectral form to unravel the empire of reason.[68] Living thought, on the other hand, maintains an open relation with this ineliminable, inexhaustible biological substrate, with the consequence, as Esposito underlines, that conflict is *constitutive* – instead of inimical – of – an always-provisional – order.[69]

A second, yet intimately connected consequence, is that, as already implied by the label of living thought, philosophy in this tradition is always already and inevitably immersed in life. There is no thought that is not also necessarily thought *in action*. Moreover, as thinking philosophically is invariably immersed in a plane of immanence where conflict is a key register, philosophy is relentlessly political philosophy: philosophy always takes sides.[70] In this respect, it is important to mark the interminably productive nature of knowledge, a consideration that is particularly central to process philosophy. From the latter perspective in fact, '[t]here is, in principle, no theoretical limit to the different lines of consideration available to yield descriptive truths about any real thing whatever'.[71] Yet for all the boundless possibilities that knowledge may offer, we need to (re-)activate a horizon of sense – articulated through a plurality of practices, forms of knowing and perspectives – with the sufficient political and ethical force that provisionally makes sense of reality and that can act as a thorough counter-force to the Anthropocene.

Averting ecocides: *religare* as an alternative epistemic imaginary

As mentioned above, we are primarily concerned with a politics and philosophy of life, one taking inevitably side in relation to the multiple processes of dis-embodying and re-embodying that operate today in the world. At this point, however, we need a frame of reference in relation to what disembodying and/or re-embodying are. This frame of reference, this new horizon of sense, we propose, is provided by *religare*.

Religare is one of the attributed etymological roots – the most popular – to religion.[72] *Religare* signifies 'to bind fast', 'to put together', 'to assemble'. We must note that, for our purposes here, *religare* bespeaks a layout, a disposition, an arrangement of the components that make up life. On this reading, 'religion' – in its most direct etymological significance – is not so much a bundle of precepts that mediates one's prosaic affairs with the supernatural as a grid or *architectura*, an ordering of the components of life in their mutually constitutive relations. *Religare* is decidedly immanent to life – not transcendent to it. It is a layout that juxtaposes the different patterns contrived by *Weltanschauungen*, policies, economic strategies and personal intentions – without ever conflating them into a unique configuration. In order to illustrate the way in which *religare* operates as an alternate horizon of sense, we offer two examples: one is related to the re-embodiment of the 'twin' concepts of freedom and liberty – crucial to liberal modernity – and the other relates to ways of re-embodying knowledge.

Religare *re-embodies freedom*

The idea of freedom and its cognate term liberty – for our purposes here the two terms can be treated as synonyms – is one of the key concepts of modern political and legal theory. Indeed, the very essence of the modern subject is linked to a cluster of intertwined concepts – namely liberty, freedom, autonomy, independence, reason, will, law and property – where freedom is central. Yet this tradition runs on an utterly disembodied basis. Against this backdrop, in this section we wish to address the question of (re-)embodying freedom so that the latter may reconnect with its richer social and ecological dimension. Again, the deployment of an etymological method will help us open up the semantic field of freedom as it reveals a rich genealogy that (re-)locates freedom within the horizon of sense conveyed by *religare*. We will draw on the work already carried out by Roberto Esposito in this regard.[73]

Modernity, observes Esposito, enacts a semantic draining of the concept of freedom that results in a progressive loss of meaning.[74] The modern concept of freedom refers in fact either to the negative sense of being free from interference, or to the positive sense of being one's own master[75] – which marks a semantic and conceptual intersection with the concept of autonomy – and more in general with the entire conceptual cluster mentioned above. Both these senses enact the epistemic framework of *solvere*. The etymology of both freedom and liberty, by contrast, offers an entirely distinctive view.

Freedom[76] in its 'germinal nucleus, alludes to a connective power that grows and develops according to its own internal law, and to an expansion or to a deployment that unites its members in a shared dimension'.[77] Freedom is, on this reading, 'both affirmative and relational'.[78] Modern freedom, on the other hand, semantically re-calibrated and reduced, *resolves* the individual subject from its other, and, through an immunitary mechanism that, we have seen in relation to Hobbes and Locke, operates as the *dipositif* aimed at protecting the individual 'from the interference of others'.[79] From this point of view then, liberty and freedom, even in their more trivial representations as freedom of choice in a market society, are a far cry from freedom explored and experienced from within the horizon of sense of *religare*. Indeed, freedom does not, from the perspective of *religare*, dis-integrate an embodiment from its situation, from its relations. Embodiments always carry within them the flow of relations for which they act as hubs, as point of materialization and of action. Freedom, that is, is a doing, a practice that is epistemically and performatively oriented in entirely different ways – dis-embodying or re-embodying – according to the frame of reference, to the horizon of sense within which it is articulated and understood.

Religare *re-embodies knowledge*

As Nietzsche already observed, the search for *logos* and truth as foundation or *Grund* is the trademark of Western metaphysics,[80] and of modern epistemology in particular.[81] This critical deconstruction of modern epistemology, already offered from a variety of critical perspectives,[82] has been however repeatedly accused of relativism.[83] In this respect, broadly postmodern reflections on modernity are often caught in what Hardt and Negri call an 'epistemological impasse' hinging on the 'opposition of the universal and the particular'.[84] What of knowledge, if no universal claim to truth can be defended? *Religare* offers a way out of this impasse through radically linking knowledge with life. Knowledge – and truth – does not disappear into a kaleidoscope of subjective positions. Knowledge is rather measured by its effect on life – either disembodying or (re-)embodying. Truth, in other words, emerges as an ethico-political category.

This resonates with Lorraine Code's notion of 'truth to'.[85] From Code's perspective, truth is a form of interpretation – rather than of verification, and is 'textured and responsive'.[86] As a form of responsible knowledge, truth becomes a way of living and a form of politics, to the extent that knowledge is cognizant of the 'multiply contestable' nature of forms of knowledge that tend to impose permanent closure on the living world.[87] This is most immediately evident in the operational performativity of law. Defining and classifying the real – a typically legal operation – produces and re-produces the world – a *certain* world – in ways that always imply the commitment to, and the reproduction of a particular epistemology.[88] This legal epistemology, Delaney further observes, by shaping and delineating the boundaries of categories such as 'nature', 'wilderness', 'body', 'mind' and 'animal', marks concretely, and sometimes violently, 'segments of the material world'.[89]

Definitions in law – as both delineations and official representations of reality – do indeed have tremendous effects on the living, human and non-human world(s). In a horizon of sense responsive to *religare*, and hence from the perspective of the framework of re-embodiments, the crucial question then is *not* whether knowledge – e.g. a certain epistemology of nature – is true, *but rather* whether it is disembodying or re-embodying, in the sense explained above. To reiterate, what matters are the *effects* of knowledge, because, as new-materialist philosophers suggest, 'matters of fact, matters of concern, and matters of care are shot through with one another'.[90]

Exploring re-embodiments by way of a collection of errancies

As in all collections, contributions offer a wide spectrum of interpretive registers of the overall theme. Yet all contributions to this collection can be folded, on our reading, within the horizon of sense that the framing of *religare* outlines. Indeed, *religare* is the epistemological fold within which all contributions arguably manoeuvre theoretically, and provides a common overarching background against which all contributions can be read.

The collection in this respect should be understood as an ensemble of explorations, of errancies, of insurrections, all tentatively carving space for re-embodying counter-trajectories. All contributions, to be sure, exhibit a theoretical congruence even if the language of re-embodiments is not always explicitly deployed – yet always conceptually at work.

This volume, as already mentioned, represents a collective – as yet unstable – weaving effort. It is strategically structured in three parts. Part one explores possible and currently underway pathways to go 'Beyond Modernity', to leave the ecocidal tendencies of the latter behind and submerge ourselves into re-embodying praxis. As one such central pathway is the retrieval of a sense of the holy, the second section of the present collection is devoted to 'The Sacred Dimension'. Here, to be sure, four authors explore the centrality of the sacred for re-embodying practices. A third section deals with 'The Legal Dimension'. The contributions in this section explore a number of legal errancies that all articulate theory as practice, and draw, tentatively or more decisively, theory from practice.

Part one, 'Beyond Modernity', comprises three essays. In 'Beyond Modernism and Postmodernism: The Narrative of the Age of Re-Embodiments', the Australian philosopher Arran Gare starts off by presenting re-embodiments as a narrative that is closely bound up with the establishment of truly democratic systems, such as those exemplified by Earth Democracy and Ecological Civilization. The narrative of re-embodiments, Gare claims, is to oust the illusory, long-standing quest of disembodiment, which has facilitated the exploitation of others. Gare establishes a direct link between the modernist and postmodernist forms of the illusion of disembodiment and that on which medieval civilization was based where the military aristocracy and the clergy, defining themselves through Neoplatonic Christianity and identifying with the eternal, despised nature, the peasantry and women. In chapter II, 'What is the Age of Re-Embodiments? Or, the Victorious Assertion of *Loci Standi* over the Barbarism of *Instrumenta Movendi*', the Catalan ecophilosopher Ruth Thomas-Pellicer shows how some binary pairs profoundly characteristic of Western metaphysics – living organisms/inert matter, theism/atheism and anthropocentrism/eco centrism – are at the root of global ecocide and follow from the disembodied and disembodying character of the former. In endeavouring to embody

Western metaphysics, Thomas-Pellicer comes up with two novel categories of knowledge: *loci standi* or 'places of secure stay' and *instrumenta movendi* or 'instruments of mobility'. These two neologisms enable a distinctive reassessment of reality. Thomas-Pellicer claims that the re-embodiment of the Western/ized world goes by way of hinging civilization on *loci standi* whilst relinquishing our dependency on *instrumenta movendi*.

This section is closed with an essay by the US ecofeminist Madronna Holden entitled 'Reclaiming Authenticity and Ethics: An Ecofeminist Vision of Re-Embodiment'. In this essay, Holden contrasts the Eurocentric worldview – which is also visible in her colonial adjuncts – with the one that follows from adopting a re-embodiment attitude. The latter, Holden clarifies, re-integrates the broken pieces of the Eurocentric vision that relentlessly translates into the divorce of mind from body, self from other and humans from nature while detaching many from both their body and those with a body, which results in the denial of ecological and social interdependence, and the privileging of some humans over others and of all humans over the natural world. On the basis of her reintegration, Holden models both personal authenticity and ethical decision-making in the context of ecological relationships.

Let us move to section two, The Sacred Dimension'. In the first chapter of this section, 'Towards a Deconstruction of Leadership and Cosmology: The Re-Embodiment of the Sacred', the Scottish ecofeminist Ali Young adopts a postmodern approach from where to confront the logic of 'hyper'-modernity and re-invent it. Young proceeds by working within the fields of leadership and cosmology. She seeks to compare two contrasting theoretical approaches to the analysis of socially constructed hierarchy involving the sacred and the profane. On the basis of two texts, Young exhibits how the divine has been disembodied and rendered transcendent, a process that proves utterly anti-ecological. Ultimately, Young argues that the ways in which we construct the sacred are of considerable importance to the potential creation of an Age of Re-Embodiments. In chapter V, 'From Enlightenment to Enchantment: Changing the Question', the London-based ecophilosopher Patrick Curry presents Immanuel Kant's and Michel Foucault's respective addresses of the question 'What is Enlightenment?', published in 1784 and 1984, not so much as hallmarks of modernity as a period but instead as a sensibility which is decidedly disenchanted. In sharp contrast, Curry proposes a non-modern (rather than pre- or post-) axio-sensibility marked by an embodied, ecocentric (non-anthropocentric, that is) and animist enchantment. To make his case, Curry draws on the work of Maurice Merleau-Ponty, Bruno Latour and Eduardo Viveiros de Castro.

In chapter VI, '(Re)embodying Which Body? Philosophical, Cross-Cultural and Personal Reflections on Corporeality', UK-based environmental anthropologist and political ecologist Sian Sullivan highlights the complexity and ambiguity intrinsic to re-embodiment, given that 'the body' is immersed in history and culture and thus invariably caught in specific regimes of truth – in the Foucauldian sense of socially constructed reality. After reviewing the great divides endemic to modernity with their attendant devastating ecological effects, Sullivan adopts a poststructuralist approach in relation to body, corporeality and embodiment and suggests that animist orientations towards knowing – from indigenous contexts to contemporary eco-paganism – refract the socionatural uniformity assumed by modernity. Finally, British embodiment theorist and ecotherapist Adrian Harris closes the section on 'The Sacred Dimension' with an essay entitled 'The Knowing Body: Eco-Paganism as an Embodying Practice'. Harris retells that eco-pagan practice awakens their practitioners from the dualistic dream that we are separate from the 'wisdom of the body'. Eco-pagan practice consists in the establishment of an intimate relationship with aspects of the natural environment in terms of spirits of place – *genius loci*, which enables eco-pagans to think with place. Harris theorizes this embodied knowing, which indeed takes him beyond ontological dualism, by partially relying on the embodied epistemology of philosopher and psychologist Gendlin.

The third part is dedicated to 'The Legal Dimension'. Here four essays offer multiple and complementary readings of re-embodiments, all taking as their starting point – implicitly or explicitly – the disembodied and disembodying character of legal modernity, whose central epistemological threshold is that established between nature and culture, and whose central construction is the (disembodied) legal subject.[91]

Alessandro Pelizzon, an Italian–Australian legal anthropologist and Gabrielle O'Shannessy, an emerging Australian legal theorist, both in close proximity with Earth Jurisprudence, start from the premise that all law is, ultimately, necessarily embodied – within institutions, networks of relations and, importantly, culture. On this basis, they deconstruct the liberal conceptualization of the autonomous self, which underpins the Western, positivist concept of the legal person at least since Locke. From the perspective of wild law, the authors re-activate an alternative set of 'philosophical roots' (from Spinoza to indigenous legal ontologies and cosmologies), as well as explore new frameworks (such as speculative realism), in order to articulate a different, *ecological* notion of self and of legal personhood.

Italian critical legal theorist Vito De Lucia, after a preliminary outline of how legal modernity is *both* ideo-ontologically[92] disembodied *and* performatively disembodying, offers a conceptually articulate framing of ecology as a 'transversal'[93] framework that effectively operationalizes the metaphor of *religare* discussed in this introduction. This transversal ecology, investing the natural, the social and the psychological, finds an immediate ally in the practices of the commons. In this sense, De Lucia tries to combine the 'thinking of theory with the doing of practices, in order to open space for articulating an ecological, re-embodying legal philosophy'.[94]

Italian socio-legal scholar Margherita Pieraccini outlines a theory of socio-ecological legal pluralism. Unlike the two main strands of legal pluralism (one centred on social facts and the other on discourses), socio-ecological pluralism takes its starting point from the embodied practices of *both* human and non-human actors, in what she calls 'materialist recuperations'.[95] Pieraccini develops thus a theoretical framework that 'pays attention to the way in which law is performatively given meanings in social-ecological settings',[96] and thus ascribes legal performativity to the multiply complex agency of humans and non-humans. Drawing on law and geography literature, Pieraccini offers a 'third way' of legal pluralism, conceptually and ecologically richer than the two main strands of legal pluralism that understand law as an exclusively social product – either a fact or a discourse. In so doing, she expands the effective meaning of legal agency. Pieraccini re-embodies law in its thick, material, spatial context, as shown through two case studies, where law is co-produced through socio-ecological 'encounters'.

The contribution of Sue Farran, an eclectic British law-in-society comparatist, explores how graffiti and guerrilla gardening, as 'new and evolving relationships between people and the natural and built environment',[97] question settled liberal boundaries between public and private and between legal and illegal, thus re-mapping the socio-legal space in urban environments. These practices, suggests Farran, signal a shift in landscape, which demands a corresponding change in lawscape. Moreover, these 'counter-revolutionary' practices re-embody legal agency, to the extent that both graffiti and guerrilla gardening – albeit in different ways and eliciting different responses from public authorities – reclaim common places *against* their disembodiment and commoditization as appropriable things; reclaim communal legalities *against* what Farran calls 'state ubiquity'; and reclaim inclusive access and use rights *against* exclusive proprietary title.

Conclusions: binding tightly our hopefully re-embodying future

We have claimed that modernity is informed by an overarching organizing epistemic captured by the rubric '*solvere*'. *Solvere* endlessly entails the dissection of the whole into multiple parts. In this light, the whole is figured out not by integrally grasping it but by separately analysing its building pieces. This was, to be sure, Galileo's innovative method, which was subsequently epitomized by Descartes's definitive split between the *res cogitans* – the human mind – and the *res extensa* – human corporeality and that of animals conceived as deprived of any valuably mindful activity.

Yet in line with the aim of the BSA workshop that gave the initial birth to the present collection, the moving of Western/ized civilization into a post-ecocidal stage entails the wholesale re-imagination of our contemporaneity. In this radical move, we need to locate with sufficient aplomb an epistemic horizon able to effect a noted quiver on *solvere*. We have claimed that this epistemic horizon is offered by the rubric '*religare*'. The most popular etymological root of religion and meaning 'to bind tightly', *religare* joins back the scattered parts; it restores the whole. *Religare*, that is to say, heals the wounds inflicted by the resolute – read this term also in an etymological sense – force of modernity.

In the wake of Nietzsche and Heidegger, we have exhibited how the exit doors of the environmentally unfriendly 'iron cage' – to now emphasize, in Weberian terms, one aspect of modernity – are conformed by all ideas, concepts and processes that disobey the modern canon in so far as they have been conceived as unorthodox by the latter. In this light, *religare* may be conceived as the chief errancy of modernity: it binds together the ecocidal dis-solutions.

The present collection is in itself a preliminary *religare* of insurrections. We hope that their reading will stimulate the expansion and further dissemination of this benign – and utterly vital – 'religion' of our hopefully re-embodying future.

Notes

1   R. Thomas-Pellicer, 'Reimagining our Sociological Contemporaneity: What is the Age of Re-Embodiments? BSA Theory Study Group Symposium Booklet, London: 16 July 2010. Available online at: www2.warwick.ac.uk/fac/soc/sociology/staff/academicstaff/gurminderkbhambra/research/bsatheorygroup/event2010/reimagining.booklet.pdf (accessed on 30th April 2015).

2   This is the case of Arran Gare's 'Beyond Modernism and Postmodernism: The Narrative of the Age of Re-Embodiments; Ruth Thomas-Pellicer's 'What is the Age of Re-Embodiments? Or, the Victorious Assertion of *Loci Standi* over the Barbarism of *Instrumenta Movendi*'; and Ali Young's 'Towards and Deconstruction of Leadership and Cosmology: The Re-Embodiment of the Sacred'.

3   De Lucia, this volume.

4   This is increasingly recognized as the task of scholarship, see e.g. M. Pieraccini, 'Reflections on the Relationship between Environmental Regulation, Human Rights and beyond – with Heidegger' in A. Grear and E. Grant (eds) *Thought, Law, Action and Rights in the Age of Environmental Crisis*, Cheltenham: Edward Elgar, 2015.

5   It is Giddens who speaks most explicitly of postmodernity as a 'more radicalised and universalised' form of modernity, A. Giddens, *The Consequences of Modernity*, Stanford, CA: Stanford University Press, 1990, p. 3. In fact, Giddens eschews altogether the term 'postmodern'.

6   And here one thinks of course of the Cartesian disembodied subject.

7   This is an increasingly important point made by critical legal scholarship in Latin American, see e.g. José Luiz Quadros de Magalhães, *O Estado Plurinacional e o Direito Internacional Moderno*, Curitiba: Juruá Editora, 2012, which speaks of modernity as entailing the 'systematic negation of diversity',

p. 15. But in the sense of a simplifying and homogenizing modernity (with particular respect to property) see also P. Grossi, *L'Europa del Diritto*, Bari: Laterza, 2001.

8   Millennium Ecosystem Assessment, *Ecosystems and Human Well-being: Synthesis*, Washington, DC: Island Press, 2005; IPCC, 2014: Summary for policymakers. In: Climate Change 2014: Impacts, Adaptation, and Vulnerability. Part A: Global and Sectoral Aspects. Contribution of Working Group II to the Fifth Assessment Report of the Intergovernmental Panel on Climate Change [C.B. Field, V.R. Barros, D.J. Dokken, K.J. Mach, M.D. Mastrandrea, T.E. Bilir, M. Chatterjee, K.L. Ebi, Y.O. Estrada, R.C. Genova, B. Girma, E.S. Kissel, A.N. Levy, S. MacCracken, P.R. Mastrandrea, and L.L. White (eds)]. Cambridge, UK and New York, NY, USA: Cambridge University Press, pp. 1–32; UNEP, *Environment for the Future We Want. GEO 5* (United Nations Environment Programme 2012); Food and Agriculture Organization of the United Nations (FAO) *The State of World Fisheries and Aquaculture. Opportunities and Challenges* (FAO 2014).

9   Ecocide is more precisely defined as 'the destruction of large areas of the natural environment especially as a result of deliberate human action'. Available online at: www.merriam-webster.com/dictionary/ecocide (accessed on 25th June 2015). For a comprehensive articulation of the concept of ecocide and its potential legal implications see e.g. P. Higgins, *Eradicating Ecocide: Laws and Governance to Stop the Destruction of the Planet*, London: Shepheard-Walwyn, 2010; see also Thomas-Pellicer and Curry, both this volume.

10   Either directly, through so-called geoengineering technologies; or indirectly, through the incentives towards technological development supposedly to come from the marketization of environmental protection. In this respect, technology transfer is central to the legal regime most closely linked to the pervasive effects of humanity on the planet, namely the climate regime. See e.g. D. Ockwell and A. Mallett (eds) *Low Carbon Technology Transfer: From Rhetoric to Reality*, Abingdon: Routledge, 2012; V. De Lucia, 'The Climate Justice Movement and the Hegemonic Discourse of Technology', in M. Dietz and H. Garrelts (eds), *Handbook of the Climate Change Movement*, Abingdon: Routledge, 2013.

11   As evident by increasing centrality of the framework of ecosystem services, and its associated markets, in the context of conservation, see the initiative called The Economics of Ecosystems and Biodiversity, a 'a global initiative focused on "making nature's values visible"', and whose 'principal objective is to mainstream the values of biodiversity and ecosystem services into decision-making at all levels'. This is achieved by helping 'decision-makers recognize the wide range of benefits provided by ecosystems and biodiversity [and] demonstrate their values in economic terms'. Available online at: www.teebweb.org/about (accessed on 25th June 2015). For a critique of the marketization of nature see e.g. S. Sullivan, 'Green Capitalism, and the Cultural Poverty of Constructing Nature as Service Provider', *Radical Anthropology 3*, 2009, 18.

12   A. Grear, 'Multi-Level Governance for Sustainability: Reflections from a Fractured Discourse – a Response to Bosselmann', in K. Bosselmann and A. Grear (eds), *New Zealand and the EU: Contested Futures: Sustainability, Governance and International Human Rights*, (Europe Institute, University of Auckland 2010) p. 73.

13   B. de Sousa Santos, *Toward a New Common Sense. Law, Science and Politics in a Paradigmatic Transition*, New York, London: Routledge, 1995.

14   Leiss draws in detail the distinction between the epistemic dimension (mastery) and the political one (domination), W. Leiss, *The Domination of Nature*, Montreal: McGill-Queens University Press, 1994.

15   M. King, 'Heidegger's Etymological Method: Discovering Being By Recovering The Richness Of The Word', *Philosophy Today*, 51:3, 2007, 278, p. 278.

16   Ibid.

17   Online Etymology Dictionary. Available online at: www.etymonline.com/index.php?term=solve&allowed_in_frame=0 (accessed on 31st March 2015).

18   B. Babich, 'On the Analytic-Continental Divide in Philosophy: Nietzsche's Lying Truth, Heidegger's Speaking Language, and Philosophy', in C.G. Prado (ed.), *A House Divided: Comparing Analytic and Continental Philosophy*, Amherst, NY: Prometheus/Humanity Books, 2003, pp. 63–103.

19   C. B. Macpherson, *The Political Theory of Possessive Individualism. Hobbes to Locke*, Oxford: Oxford University Press, (1962) 2011, pp. 30ff. But see also H. Jonas, The Scientific and Technological Revolutions, *Philosophy Today*, 15:2 (1971: Summer) 76.

20   Synthetic biology is expected to add 'novel forms of biogenic sedimentation' to the already effected 'biotic, sedimentary and geochemic change' characterizing the Anthropocene, A. Ginsberg, J. Calvert, P. Schyffer, A. Elfick and D. Endy, *Synthetic Aesthetics: Investigating Synthetic Biology's Designs on Nature*, Cambridge, MA: MIT Press, 2014, p. 197.

21   The fact that metaphors are characterized by unsteady, tentative, tensive – as opposed to firm, solid, definite – truths is the great insight advanced by P. Ricour. The statement, say, 'Achilles is a lion' is only true in a certain, shaky way, namely, in the sense that Achilles' strength *evokes* that of a lion. It is untrue in the actual sense that Achilles' strength be fully equivalent to that of a lion. P. Curry, 'Radical Metaphor: or why Place, Nature and Narrative are each other but aren't themselves', *EarthLines*, 2013, (6), pp. 35–8. Interestingly, this fragile, nuanced, subtle manner of relating meaning is the one that sustains not only general communication but also scientific discourse.

22   See Santos op. cit. for the outline of the isomorphic relation between the two domains.

23   P. Crutzen, 'Geology of mankind', *Nature*, 2002, 415 (6867), p. 23.

24   Ibid; J. Zalasiewicz, M. Williams, W. Steffen and P. Crutzen, 'The New World of the Anthropocene', *Environmental Science and Technology*, 2010, 44 (7), pp. 2228–31.

25   S. Lewis and M. Maslin 'Defining the Anthropocene', *Nature*, 2015, 519, pp. 171–180.

26   Ibid, p. 171.

27   Such as *Anthropocence*, published by Elsevier. Available online at: www.journals.elsevier.com/anthropocene/ (accessed on 30th June 2015); *The Anthropocene Review*, published by Sage. Available online at: http://anr.sagepub.com/ (accessed on 30th June 2015); *Elementa: Science of the Anthropocene*, published by BioOne. Available online at: www.elementascience.org (accessed on 30th June 2015).

28   See, e.g. D. Botkin, *Discordant Harmonies: A New Ecology for the Twenty-first*

Century, Oxford University Press, 1st edition, 1990; M. Marvier, 'New Conservation is True Conservation', *Conservation Biology*, 2013, 28 (1) pp. 1–3; M. Marvier and P. Kareiva, 'The Evidence and Values Underlying "New Conservation"', *Trends in Ecology and Evolution*, 2014, 29 (3) pp. 131–2.

29   Botkin, op. cit., p. 193.

30   See, among an increasing scholarship, the papers presented at The Thousand Names of Gaia: From the Anthropocene to the Age of the Earth, Rio de Janeiro, 15–19 September 2014. Available online at: https://thethousandnamesofgaia.wordpress.com/the-conferences-texts/ (accessed on 30th April 2015). See also, e.g. L. Kotzé 'Human Rights and the Environment in the Anthropocene' *The Anthropocene Review*, 2014, 1 (3) pp. 252–75; D. Vidas, 'The Anthropocene and the International Law of the Sea', *Philosophical Transactions of the Royal Society A*, 2011, 369, pp. 909–25.

31   Kotzé, op. cit., p. 252.

32   Grear and Grant, op. cit.

33   On disembodiment as a (meta)trajectory see De Lucia, this volume.

34   A. Grear, 'The Vulnerable Living Order: Human Rights and the Environment in a Critical and Philosophical Perspective', *Journal of Human Rights and the Environment*, 2011, 2 (1), pp. 23–44.

35   G. Capograssi, Riflessioni sull'Autorità e la Sua Crisi, (1921), in G. Capograssi, *Opere*, Milano: Giuffrè, 1959, 1, p. 331, [our translation].

36   R. Esposito, *Bios: Biopolitics and Philosophy*, Minneapolis, MN: Minnesota University Press, 2007, p. 45.

37   Ibid., p. 61.

38   Ibid., p. 61.

39   Ibid., p. 61.

40   E. Pashukanis, *Law and Marxism: A General Theory*, London: Pluto Press, 1989.

41   B. Weston and D. Bollier, *Green Governance. Ecological Survival, Human Rights, and the Law of the Commons*, Cambridge: Cambridge University Press, 2013.

42   See e.g. M. Tallacchini, *Diritto per la Natura. Ecologia e Filosofia del Diritto*, Milano: Giappichelli Editore, 1996; C. Cullinan, *Wild Law: A Manifesto for Earth Justice*, Siber Ink: South Africa, 2002; P. Burdon, *Earth Jurisprudence: Private Property and the Environment*, London: Glasshouse/Routledge 2014.

43   As already noted by Hegel, see D Loick, '"Expression of Contempt": Hegel's Critique of Legal Freedom', *Law and Critique*, 2015, 26, pp. 189–206.

44   Loick, op. cit., p. 193.

45   Ibid., p. 193.

46   See e.g. A. Grear, 'Deconstructing Anthropos: A Critical Legal Reflection on "Anthropocentric" Law and Anthropocene "Humanity"', *Law and Critique*, 2015. Available online at: http://dx.doi.org/10.1007/s10978-015-9161-0 (accessed on 15th July 2015).

47   It is exactly for this reason that Anna Grear speaks of '*quasi*-disembodiment', A. Grear, *Rethinking Human Rights. Facing the Challenge of Corporate Legal Humanity*, London: Palgrave/McMillan, 2009.

48   Loick, op. cit., p. 193.

49   Grear 2009, op. cit. Grear speaks in this respect, paradoxically, of 'corporate legal humanity'.

50   Loick, op. cit., p. 193.

51   G. Vattimo, *The End of Modernity: Nihilism and Hermeneutics in Postmodern Culture*, Baltimore, MD: The Johns Hopkins University Press, 1991 [1988]; trans. and with an introduction by Jon R. Snyder, p. 166.

52   See A. Williams and N. Srnicek, '#ACCELERATE MANIFESTO for an Accelerationist Politics'. Available online at: www.criticallegalthinking.com/2013/05/14/accelerate-manifesto-for-an-accelerationist-politics/ (accessed on 26th May 2015). For a critique, see A. Negri, 'Some Reflections on the #ACCELERATE MANIFESTO'. Available online at: http://criticallegalthinking.com/2014/02/26/reflections-accelerate-manifesto/ (accessed on 26th May 2015).

53   Vattimo, op. cit., p. 174.

54   Or what in Foucauldian terms would be called insurrections, a frame indeed deployed in one of the contributions to this collection, see De Lucia this volume.

55   Cf. D. Coole and S. Frost (eds), *New Materialisms: Ontology, Agency and Politics*, Durham and London, Duke University Press, 2010.

56   Ibid., p. 7.

57   Ibid.

58   Ibid.

59   De Lucia, this volume.

60   The identification of what Nietszche sees as the unconditional, objectifying authorities, logos and truth, as the local effects of a particular system has been conceived by Arkady Plotnitsky as the Nietzschean 'revolution'. This major scholarly input rightly attributed to Nietzsche signifies that authorities and subordinations emerge and impose themselves all the time. A. Plotnitsky, *Reconfigurations: Critical Theory and General Economy*, University Press of Florida, 1993, pp. 152–3.

61   F. Guattari, *The Three Ecologies*, London and New York, Continuum, 2008.

62   G. Bataille, *On Nietzsche*, London and New York: Continuum, 2004. Translated by Bruce Boone; introduction by Sylvère Lotringer, p. ix.

63   Esposito uses the label 'living thought' to describe an 'Italian' tradition, but that is inevitably connected, with multiple lines, with the rebel philosophies (such as, for example, Spinoza's or, especially, Nietzsche's and Heidegger's) that traverse modernity, R. Esposito, *Pensiero Vivente. Storia e Attualità della Filosofia Italiana*, Milano: Einaudi, 2012.

64   Whenever genealogy is invoked, one is immediately drawn to Foucault and to Nietzsche (see e.g. M. Foucault, 'Nietzsche, Genealogy, History', in M. Foucault, *Language, Counter-Memory, Practice: Selected Essays and Interviews*. Itacha, NY: Cornell University Press, 1977). The link exists, but *post hoc*, in the sense that Esposito deploys the language of genealogy to give coherence to a long political philosophical tradition (going from Machiavelli to Agamben, Negri and Esposito himself) that used different ways to articulate the same fundamental concept of history.

65   Esposito 2012, op. cit., p. 24.

66   T. Hobbes, *Leviathan*, 1651, eBooks@Adelaide The University of Adelaide Library, Chapter 13.

67   Ibid.

68   A return whose most intense and violent example – an example that has been interpreted as the necessary horizon of modernity as well as its paradigm, by e.g. Agamben – is the Nazi camp. See e.g. G. Agamben, *HOMO SACER*

Sovereign Power and Bare Life, Stanford, CA: Stanford University Press, 1998; Esposito 2007, op. cit.

69 Esposito 2012, op. cit., p. 27.

70 This is most immediately apparent in the philosophy of Machiavelli and Gramsci, both of which are discussed in Esposito 2012a, op. cit.

71 N. Rescher, *Process Metaphysics: An Introduction To Process Philosophy*, New York: State University of New York, 1996, p. 130; see also J. Seibt, 'Process Philosophy', Standford Encyclopedia of Philosophy, 2012. Available online at: http://plato. stanford.edu/entries/process-philosophy/ (accessed on 30th April 2015).

72 F. Capra, 'The Role of Physics in the Current Change of Paradigms', in R.F. Kitchener (ed.), *The World View of Contemporary Physics: Does it Need a New Metaphysics?*, New York: State University of New York Press, 1988, pp. 144–55; pp. 145–6. The alternative etymology of religion, *re-ligere*, indicates as its primary sematic referent that of care, a metaphorical transposition of the literal meaning of discern, choose, select implicated in *legere*. So, religion, indicates from this point of view a continuous engagement premised on practices of care, see O. Panigiani, *Vocabolario Etimologico*, Milano: Albrighi & Segati, 1907, 'religione'. Available online at: http://etimo.it/?term=religione &find=Cerca (accessed on 30th April 2015).

73 Esposito 2007, op. cit.

74 Ibid., p. 70.

75 Isaiah Berlin, as discussed in ibid., p. 71.

76 As well as that of liberty, through a different yet congruent etymological route, ibid., p. 70.

77 Ibid., p. 70.

78 Ibid., p. 71.

79 Esposito 2007, op. cit., p. 72.

80 Plotnitsky, op. cit.; Vattimo, op. cit.

81 R. Rorty, *Philosophy and the Mirror of Nature*, Princeton, NJ: Princeton University Press, 1980.

82 Including post-colonialism, race, and feminist theory. Cf. S. Harding, *Sciences from Below: Feminisms, Postcolonialities, and Modernities*, Durham and London: Duke University Press, 2008.

83 J. Habermas, 'The Entry into Postmodernity: Nietzsche as a Turning Point', in *The Philosophical Discourse of Modernity: Twelve Lectures*, Cambridge, UK: Polity Press, 1987; translated by Fredrick Lawrence, pp. 83–105.

84 Hardt and Negri, op. cit., p. 120.

85 L. Code, *Ecological Thinking: The Politics of Epistemic Location*, Oxford: OUP, 2006, p. 7; Code also draws from Wittgenstein.

86 Ibid., p. 7. See De Lucia this volume for a more detailed exploration of this theme in relation to the frame of this collection.

87 Code op. cit., p. 50.

88 See C. Grzegorczyk, 'Le concept de bien juridique: l'impossible définition?', *Archives de philosophie du droit, Les biens et les choses*, 1979, 24, p. 259. See also, K. Tuori, *Critical Legal Positivism*, Farnham, Surrey: Ashgate, Applied Legal Philosophy series, 2002.

89 D. Delaney, 'Making nature/Marking Humans: Law as a site of (Cultural) Production', *Annals of the Association of American Geographers*, 2004, 9 (3) p. 489.

90   R Dolphijn and I van der Tuin, *New Materialism: Interviews & Cartographies*, Ann Harbor, Michigan: Open Humanity Press, 2012, p. 69. Similarly, B. Latour, 'An Attempt at a "Compositionist Manifesto"', *New Literary History*, 2010, 41, pp. 471–90.

91   Thus e.g. A. Grear, *Redirecting Human Rights. Facing the Challenge of Corporate Legal Humanity*, London: Palgrave/McMillan, 2009.

92   This is a coinage that wishes to indicate the complicit relations between an ontological horizon and the ideology that appropriates it and 'puts it to work' towards a particular project, see De Lucia this volume.

93   Drawing on F. Guattari, *The Three Ecologies*, London: Continuum, 2008.

94   De Lucia, this volume.

95   Pieraccini, this volume.

96   Pieraccini, this volume.

97   Farran, this volume.

Part I

Beyond modernity

Chapter 1

Beyond modernism and postmodernism

The narrative of the age of re-embodiments

Arran Gare

Modernity/postmodernity or the grand narrative of disembodiment

The quest for re-embodiments is a reaction against a culture that has pretended to become progressively disembodied, free of the constraints of embodied existence. It is this pretence of disembodiment that has enabled some segments of society to engage in a plethora of activities – ore mining in third world countries, development of bureaucracies with global reach, air travel, mass consumerism – that are disembodying third parties, stripping their communities of their own embodied forms. This process, which Susan George described in *How the Other Half Dies: The Real Reason for World Hunger*,[1] has accelerated with climate destabilization and wars to control oil and other natural resources. In Iraq, at least 800,000 people have been disembodied so far. The greatest obstacle to the quest for re-embodiments, I will suggest, is the entrenched, barely conscious yet powerful grand narrative of disembodiment that sanctions and normalizes these disembodying activities that surreptitiously has dominated the modern/postmodern world. But the grand narrative of disembodiment is not named as such. Rather, it is called modernism, or postmodernism. As will be suggested later, without appreciating it, proponents of modernism and postmodernism have embraced and transmogrified the Neo-Platonic Christian quest for disembodiment so completely that they are unaware of it. The quest by ruling elites to leave behind the changing, sensible world and to aspire to what is eternal, the realm of money, has characterized Modernism. As Karl Marx observed: 'The cult of money has its asceticism, its self-denial, its self-sacrifice – economy and frugality, contempt for mundane, temporal and fleeting pleasures; the chase after the *eternal* treasure. Hence the connection between English Puritanism, or also Dutch Protestantism, and money making.'[2] I will suggest that this cult of money results in a quest to overcome and leave behind the constraints of the material world. Deconstructive postmodernists, purportedly leaving behind modernity, claim that there is no extra-text, there exists nothing but an endless play of signifiers that are not only disconnected from any base in reality, but the idea of such a base is nothing but an illusion created by this play of signifiers.[3] These postmodernists are even more completely enmeshed in the grand narrative of disembodiment than the modernists. This grand narrative of disembodiment is now embodied in people's habitus, in their whole way of life, in their institutions and, most importantly, in the ends they aspire to. Progress, identified with economic growth, is seen to be moving towards a dematerialized economy. Robert Solow, a Nobel laureate in economics, argued that through substitution of produced capital for natural capital, the market could generate permanent sustainability. 'If it is very easy to substitute other factors for natural resources' he argued, 'then there is, in principle, no "problem". The world can, in effect, get along without natural resources, so exhaustion is just an event, not a catastrophe.'[4]

The drive for disembodiment is most clearly evident in the development of electronic media and the achievements of those people who are regarded as most successful in the modern world, the new globalized class of super-wealthy managers of transnational corporations and financial institutions who, along with economic advisors to governments who legitimate their dominance, now control not only most of the world's economy but also its politics. As Zygmunt Bauman observed:

> Elites travel in space, and travel faster than ever before, but the spread and density of the power web they weave is not dependent on that travel. Thanks to the new "body-less-ness" of power in its mainly financial form, the power-holders become truly ex-territorial even if, bodily, they happen to stay "in place."[5]

The global web of information and communication has annulled temporal/spatial distance for this new class, apparently emancipating them from environmental constraints.

While this class extends its apparently disembodied power around the globe, 'others watch helplessly as the locality they inhabit moves out from under their feet.'[6] Their helplessness is increased because many of them, particularly but not only in the technologically advanced countries, have been seduced by the quest for disembodiment. As Susan Greenfield observed of the effects on the new digital technologies, young people are no longer driven by desire to fulfill themselves as embodied subjects in the context of their natural and social communities, but by the quest to participate in fantasy worlds in cyberspace.[7] This is the culmination of a trend. To begin with, this was a fantasy world conjured up and portrayed by advertisers (who have replaced artists as educators of feeling). As Jean Baudrillard realized, people had become consumers of signs, which gain their value through relation to other signs, rather than products. This allows the monetary sign 'to escape into infinite speculation, beyond all reference to a real of production.'[8] As investors have moved from backing companies that produce things to speculating in fictitious capital associated with asset inflation and increasingly abstract investment products, consumers have graduated from the fantasy world of advertising to the high-definition fantasy worlds on the internet, culminating in Second Life and posthumanism.[9] As Baudrillard observed, the simulacra has replaced reality:

> It is the generation of models of a real without origin or reality: a hyperreal. The territory no longer precedes the map, nor survives it. Henceforth it is the map which precedes the territory [...] It is the real, and not the map, whose vestiges subsist here and there, in the deserts which are no longer those of the Empire, but our own.[10]

The simulated scenes of Second Life are the consumer equivalent of the optimistic econometric models of the economy produced by neo-classical economists.[11]

The different value ascribed to embodied and disembodied practices reflects the de-valorization of embodied and the valorization of the disembodied. It is those working closest to the Earth, the growers of food, who are most likely to starve or be driven to suicide through poverty.[12] Increasingly, the poor of the world are suffering the effects of land degradation and pollution, including global climate destabilization. Forced off the land in Brazil, India or China, those who work in factories also live in poverty. According to my Indian students, in India, factory accidents killing hundreds of people are so common they are not even reported. In China, factory workers work twelve hours per day or more, seven days a week, often living in dormitories.[13] In the deindustrialized countries of the West most people who are still employed, the vast majority of them living in cities, work in the service industries.[14] These same people are able to consume exotic and out of season fruits, vegetables and beverages along with seafoods and manufactured goods sent to them from all corners of the globe, which also supply minerals and oil and that are now tourist destinations and, in some cases, sources of body organs. Even here, the more disembodied workers' practices are the more they are valued. People working at the 'coalface', such as foot soldiers, waiters, writers or tutors in universities, are least well paid and least respected. The respected soldiers are the high-tech military personnel who kill people from their offices using drones and missiles that allow them to see on their screens the expressions on the faces of the targeted just before they die. The new profession of 'human resources' has become a growth industry as business enterprises, backed by government policies, work to reduce the costs of employing people who actually dirty their hands with labour. Even more lucrative careers are found in marketing, public relations and other areas associated with the production of simulacra. Managers, who have least to do with the material world, are the most exalted. The pinnacle in this system is finance, almost completely disembodied. In 2007, the financial sector in the USA gained 41 per cent of after-tax profits, up from five per cent in 1982.[15] So exalted are the members of the financial sector that when they created a global economic crisis that began in 2007 they were rewarded with huge government bail-outs, allowing them to boost their already stratospheric incomes. A grand narrative of re-embodiments is now uniting the struggle against all these attempts to deny our embodied condition.

The narrative of disembodiment and the history of macroparasitism

One of the most important tasks of a grand narrative of re-embodiments is to identify, comprehend, encompass and then supersede the grand narrative of disembodiment. It should be clear from the present state of culture that the quest for disembodiment is intimately related to what the American historian William McNeill called 'macroparasitism', people living off the produce and services of others.[16] The quest for disembodiment by macroparasites appears to have been a recurring, although not universal, feature of civilizations since they began, and it is through this quest that they have defined their superiority, legitimated their rule and justified their exploitation and oppression of those who they have exploited.

The prototypical examples of this were the pharaohs of Ancient Egypt who claimed to be descended from Ra, the sun god and to represent the gods on Earth. Their whole lives were devoted to preparing themselves to be mummified and placed in pyramids for an afterlife. When the body died, parts of its soul known as *ka* (body double) and the *ba* (personality) would go to the Kingdom of the Dead, with pharaohs retaining their divine status after death.[17] Gradually, this divinity became the aspiration of 'commoners', that is, the ruling elite, and the aspiration was central to their domination of the rest of society. Without being in any way influenced by the Egyptians, the ruling elites of Maya also portrayed themselves as representatives of the gods, and constructing monuments in the form of pyramids to these gods to symbolize their superiority to ordinary mortals, hastened the ecological collapse of Mayan civilization. Such construction intensified as the rural population, who were supporting the elites, was reaching the limits of its capacity to produce food.[18]

There were similar patterns elsewhere. In the Epic of Gilgamesh the Sumerian king of the First Dynasty of Urak, who reigned circa 2500 BC and was deemed to be two-thirds god and one-third man, aspired to immortality. He failed, having to reconcile himself to his mortality.[19] The Indians were more successful, coming to believe in the soul's immortality after the death of the body. Originally, only holy men who withdrew from public life believed that the soul was immortal, but rulers who used this to legitimate their power embraced their beliefs. The third emperor of India's Mauryan dynasty, Ashoka, who, in the third century BC, conquered and ruled most of what is modern-day India, converted to Buddhism and then sent out missionaries to proselytize this doctrine. While Ashoka's practices were benign, ruling elites in other Buddhist societies used this belief to justify a hierarchical social structure and even slavery. Belief in a potentially immortal soul spread to the Middle East and Europe, influencing both Hebrews and Greeks before crystallizing in the belief systems of Christians and Moslems.

McNeill showed how microparasitism in the form of plagues has played a major role in history,[20] then argued that macroparasitism has played a similar role.[21] The Bronze Age civilizations initially were associated with population growth, but collapsed some three thousand years ago as macroparasites over-exploited workers, peasants and the land.[22] Subsequently, this process has been repeated on larger scales. For instance, towards the end of the ninth century, the Abbasid Caliphate of Mesopotamia, which up until then was one of the greatest civilizations the world had known, collapsed. Increasing taxes on farmers and neglect of their conditions by the ruling elite oriented towards higher ends destroyed agriculture, leaving a region of empty desolation, tangled dunes and rubble strewn mounds of former settlements. Allen, Tainter and Hoekstra wrote of this: 'The occupied area had shrunk by 94 percent by the eleventh century. Population dropped to the lowest level in five millennia. Urban life in 10,000 square kilometers of the Mesopotamian heartland was eliminated for centuries.'[23] The same tendency operated in Europe. The decline of the Western Roman Empire was largely due to environmental destruction.[24] In the context of this decline the emperor Constantine embraced a form of Christianity, which, synthesizing Neo-Platonic and Hebraic thought, denied significance to embodied life and extolled a life devoted to the eternal.[25]

After the collapse of the Roman Empire, this Neo-Platonic Christian narrative provided the foundation for medieval feudalism. The church, which had been supported by emperors to facilitate their control over their empires, later succeeded in subordinating the emperors to the church through the church's promotion of the narrative of disembodiment. Peasants were treated as the lowest form of humanity, scarcely above beasts in the hierarchy of being.[26] From the twelfth century onwards Europe was almost continually at war, most of it directed to domination and blessed by the church. This involved increasingly large segments of the population, and became ever more burdensome. Another major collapse of this civilization would almost certainly have occurred if Europeans had not developed empires. Europeans used the wealth from the New World to sustain their competition, further developing their sea power and empires in the process.[27]

The dialectic between the narratives of embodiment and disembodiment in modernity

In reaction to such narratives of escape from temporality and embodiment, people have embraced and even celebrated the finitude of embodied existence. The dialectic between the narratives of disembodiment and re-embodiments can be traced through the history of civilizations, and is best understood in relation to forms of organization. With greater social complexity and greater specialization, coordination of people and actions became more problematic. As Richard Norgaard noted, there are three ways in which large numbers of people can coordinate their activities in complex societies: through bureaucracies, through markets and through institutionally supported democratic processes.[28] Each of these is capable of innovation, and each has co-evolved with the others and with civilizations. While traditional hunter gatherer societies were democracies with joint action achieved through discussion to reach a consensus, this became too time consuming with civilizations that originally were characterized by a specialized military caste, slavery and, in some instances, feudal hierarchies, and the development of bureaucracies to administer these.[29] Markets developed within the order created by these civilizations. While democracies re-emerged in civilizations, to deal with the far greater complexities of civilization, such democracies required a high level of education and cultural development among citizens, imparting a strong sense of responsibility for and loyalty to their communities and a commitment to the common good, and the development of formal procedures for debating issues and decision-making.

Efforts to revive democracy have begun as challenges to concentrated forms of power, and in the process, have challenged the celebration of disembodiment. In response, those striving to concentrate power have deified new forms of disembodiment.

As Mikhail Bakhtin has shown, peasants challenged the structure of medieval society by attacking the petrified seriousness of their masters, subverting the content of medieval ideology: asceticism, providentialism, sin, atonement and suffering associated with fear, religious awe and humility, all in the service of an oppressive and intimidating ruling class. In reaction to this ideology, peasants developed a tradition of laughter: the carnival, the parody, buffoonery and the grotesque, opening people to the laughing aspect of reality with its unfinished and open character, with the joy of embodiment, change and renewal.[30] The association between decay and creativity, symbolized by a very old pregnant woman, was celebrated.

Neo-Platonic Christianity was less radically but more comprehensively criticized with the rediscovery and revival of Aristotle's philosophy. The tradition of peasant opposition to ruling elites and Aristotelian philosophy helped pave the way for the revival and defence of republicanism and democracy in the Renaissance. The Renaissance was associated with the development of the humanities, civic humanism and the flowering of the arts and nature enthusiasm, each celebrating embodied existence, while concerned to foster the abilities of people (although excluding the poor and women) to govern themselves.

The subsequent quest for democracy has been associated with alliances with either the market against bureaucracy and feudal relations, or bureaucracy against the market. The revival of the quest for democracy in Renaissance Italy was undertaken against feudalism and the bureaucracies of the church and emperors (often playing these off against each other) and tended to foster markets in their struggle. In later centuries, proponents of democracy attempted to make bureaucracies serve them to maintain the conditions for democracy against the forces of the market. Insofar as markets or bureaucracies have been aligned with democracy, their narratives of disembodiment have been muted. Due to these shifting alliances, the conflict between the narratives of disembodiment and re-embodiments has been confused. However, the fundamental opposition between the narrative of embodiment associated with democracy and the drive for disembodied power associated with those promoting bureaucratic control and the subordination of people to markets can be discerned not only in the Renaissance quest for democracy, but also in the reaction to this.

The original alliance of bureaucracies and markets against democracy was associated with the counter-Renaissance, that is, the scientific revolution of the seventeenth and early-eighteenth centuries, with Mersenne, Gassendi, Descartes, Hobbes, Boyle, Newton and Locke being the leading figures. Opposing the ideas of the civic humanists and nature enthusiasts, such figures argued for atomism, mechanism, objectivism and universalism, a world that could be understood through mathematical models and thereby made predictable and controllable. This scientific materialist view of nature reduced the human body to a machine, treating the mind as either a decontextualised substance, only contingently related to a body, or as an epiphenomenon, with humans conceived as machines driven by appetites and aversions (pleasures and pains). The first view was defended by Descartes and Kant; the second by Hobbes and Locke. At the core of this world-view is the Pythagorean belief, revived by Galileo, that only that which can be measured and quantified and can be manipulated according to simple recursive procedures – such as counting – is truly real; all else is subjective and epiphenomenal.[31] Value is first reduced to subjective experience, but then re-objectified through the disembodied realm of quantifiable monetary exchange value.

The peculiarity of this conception of physical existence (as Immanuel Kant pointed out) is that while the domain of mathematics and its operations lie entirely with the inner, private, subjective realm, this physical domain is considered the most objective of realms and the physical world is only acknowledged as real insofar as it can be made to conform to this subjective realm.[32] Scientific materialism is a form of materialism that denies any reality to the physical as it was originally understood; that is, as Aristotle characterized the physical as that which has the source of acting within itself.[33] And it denies any intrinsic meaning to nature, including the human body. The material world is conceived of as inert matter, moving according to immutable laws, mindlessly, endlessly and meaninglessly. Human existence as incarnate consciousness becomes unintelligible. As R.D. Laing wrote, 'Galileo's program offers us a dead world: Out go sight, sound, taste, touch, and smell, and along with them have gone esthetic and ethical sensibility, values, quality, soul, consciousness, spirit. Experience as such is cast out of the realm of scientific discourse.'[34] But at the same time, disembodied subjects insofar as they are economic actors engaged in monetary exchanges are deified as the absolute reference points for knowledge and value.

Stephen Toulmin revealed the hidden agenda of this cosmology – and the grand narrative it supports. The agenda was to make everything, including other people, measurable and predictable.[35] This agenda was made explicit by Henri de Saint Simon who proposed a society controlled by a 'Council of Newton', and as such it was embraced by Lenin and his successors who, turning Marx on his head, called for 'the transformation of the whole state economic mechanism into a single huge machine'.[36] However, it has been implicit in the whole project of modernity and has been associated not only with mobilization of people for economic growth, but also for war and imperialism, with progress defined ultimately in terms of the growth of income and wealth defined through money. By 1914 Europeans and their colonies controlled 84 per cent of the earth's surface, supporting an ever-more complex economic, administrative and military machine.[37] This civilization has generated complex technological, organizational and political innovations, along with systems of education and research facilitating further innovation. This complex organization, centred in cities, has been made possible by a global imperialist system and the exploitation of fossil fuels, allowing an almost complete dissociation of cultural evolution from ecological constraints.[38] This was modernity. Those whose lives are most ecologically destructive are dissociated from nature, almost blind both to the limits of the ecosystems of which they are part and which have made human life possible, and to the ecological destruction they are causing.

This whole development has been opposed by proponents of democracy, and in challenging this drive for domination, they have celebrated the particular, the mutable and the embodied. In opposition to the abstractions of science and economics, they have exalted unique individuals and their situations through the arts and the humanities. The early Romantics who opposed atomism and utilitarianism and defended democracy amplified the celebration of life in the art of Leonardo Da Vinci and then Rembrandt, along with the revival of history. They also opposed imperialism and the denigration of non-Europeans and celebrated cultural diversity. Notably among these were Rousseau and J.G. Herder. Herder rejected mechanistic thinking in science and made life, feeling and embodied action central to his whole philosophy.[39] Herder's ideas were taken up and further developed by later philosophers who placed the arts and the humanities at the centre of education.

Despite the power of ruling elites, the proponents of democracy were able to advance by exploiting the division between bureaucracies and markets. Thomas Jefferson, committed to democracy and the education required for this, supported markets for small farmers and artisans but opposed bureaucracies as hangovers from feudalism and aristocratic privilege serving wealthy elites.[40] As markets became increasingly oppressive, proponents of democracy sought to utilize bureaucracies to augment democracy. In France, after the revolution, government and its officials were reconceived as servants of the people, with the bureaucracy transformed into a civil service.[41] In Britain in 1853 the British Prime Minister, Lord Aberdeen, had a scheme drawn up to reorganize the civil service. This was the Northcote-Trevelyn Report, which had a revolutionary influence on civil services all around the world, including the USA. It recommended recruitment through the Chinese practice of open competitive examination in place of patronage and the elevation of 'generalist' education in the humanities against technical education.[42] This was to be in the service of democracy, understood as the accountability of governments and their officials or representatives to the people through elected representatives. In the twentieth century this was the form of the civil service around which the welfare state was built, providing people with the economic security required for them to fulfil their duties as citizens of democracies. Others, such as A.V. Dicey, Max Weber and Friedrich von Hayek, opposed the growth of the civil service as producing a new collectivism threatening the freedom of individuals; however, their arguments were also made in the name of democracy, or at least, were not opposed to it.[43] All these developments were associated with a growing respect for the particularities of people's embodied existence.

The alliance of markets and bureaucracy against democracy in the postmodern world

In retrospect it can be seen that there were strong propensities to undermine democracy in both bureaucracies and in markets, which became more pronounced as they expanded. Once the means of production cease to be owned by the workers, markets tend to concentrate wealth and power, impoverishing people, creating insecurity and indebtedness and fostering avarice and egocentric greed. Bureaucracies have a strong tendency to emphasize formal and instrumental rationality over substantive rationality, and through increasing regulation, to expand their control over, while at the same time insulating themselves from those whom originally they were designed to serve.[44] Markets and bureaucracies are not inherently opposed to each other. With the development of large-scale industry, the proletarianization of craftsmen and professional workers, the growth of corporations and financial institutions and their evolution into massive transnational organizations, businesses themselves became highly bureaucratized. This has been associated with the development of scientific management oriented to concentrating power, knowledge and decision-making in the hands of managers. As bureaucracies, transnational corporations not only operate in the market but deploy their bureaucratic power to extend markets through advertising, public relations, control of the mass media and to control of government policies, politicians and political parties. While dismantling the welfare state, they have transformed and expanded government bureaucracies into instruments for more effectively controlling people.

This has resulted in a fusion of bureaucracies and markets, finance and politics. The bonding of big corporations, banks and government has produced what John Perkins characterized as 'corporatocracy'.[45] Corporatocracy began with the fostering of transnational corporations by governments of technologically advanced countries (mainly the USA) and by the International Monetary Fund (IMF) and World Bank to economically and politically dominate and then suck the wealth out of technologically undeveloped countries, creating comprador elites and corrupt governments to serve their purposes.[46] Then in the 1970s and 80s transnational corporations and their allies succeeded in capturing control of governments in the technologically advanced countries.[47] Trade barriers and barriers to the movement of capital were eliminated and public institutions and civil services were privatized or reorganized to function like businesses. A new dimension to the economy was added by eliminating secure, full-time work, and by outsourcing and by forcing employees to act as entrepreneurs continually having to sell themselves. The outcome has been the creation of predator states with political parties and government bureaucracies run as businesses, with managers moving between political, government, business and financial bureaucracies, plundering public wealth, effecting massive redistributions of wealth and income to the wealthy and disempowering the general population.[48] Democracy has been 'managed', that is, effectively eliminated and replaced with what Sheldon Wolin described as 'inverted totalitarianism'.[49]

The true nature of corporatocracy is most clearly revealed in the collusion of governments and transnational agribusiness corporations to control food production, with corporations granted patents on genes of traditional crops developed by farmers over centuries, and ordinary farmers punished for trading their seeds. As Vandana Shiva pointed out: 'Corporate globalization is leading to food fascism – threatening the freedom of farmers and consumers and destroying the ecological, economic, and cultural foundation of food and agriculture'.[50] Effectively, a world order is being created fulfilling the project of transforming the entire world into one giant economic machine, eliminating democracy in practice, if not in name, and denying any significance to nature, or to people, other than as a means to generate profits defined in monetary terms.

Although the project of understanding the world mechanistically and thereby rendering democracy inconceivable was clearly articulated by Hobbes, it was an aspiration rather than something completed.[51] The project of understanding the world mechanistically with its hidden agenda advanced on a number of fronts. In the natural sciences, it advanced by extending mechanistic thinking from physics to chemistry and then to the life and human sciences. In the human sciences, it has been advanced principally through the development of economics, culminating in Walrasian neo-classical economics purportedly capturing the entire economy in a set of equations that prove their efficiency. This was the pinnacle of the denial of reality to anything that could not be quantified. More recently this programme has been vigorously pursued in psychology. The application of mathematical techniques to logic, which in turn provided the means to develop information processing technology based on binary arithmetic and Boolean logic, including computers and the Internet, has brought this project far nearer to completion.[52] These developments have virtually forced a disembodying set of practices in everyday life, making plausible Hobbes' claim that humans are machines, that reasoning is nothing but adding and subtracting, that science is nothing but a means to control the world in the service of satisfying people's appetites and that the arts and humanities are merely forms of entertainment. Information technology has also provided the means to organize societies on these assumptions.

With the new corporatocracy promoting neoliberalism, economists have taken over from civil servants who had a generalist, humanist education and have set about interpreting and evaluating the entire world through the categories of the market. As part of this process, universities and institutions of research are being transformed into transnational business corporations, with arts and science faculties being displaced as the core of universities by business faculties, which have also absorbed economics faculties.[53] The humanities and humanistic human sciences are being eliminated. The failure of mainstream neo-classical economists to predict major events such as the 2007 financial crisis (a crisis largely due to governments following those economists' policy recommendations), along with their imperviousness to criticism, should dispel the illusion that these economists are social scientists devoted to understanding society. As Robert Nelson has argued, economists 'are more like theologians' serving 'as the priesthood of a modern secular religion of economic progress'.[54] That is, neo-classical economics, reminiscent of Christian Neo-Platonism, has become a new dogmatic theology, and it is this theology that has worshipped and defined everything through the increasingly disembodied realm of money, now largely integrated with information processing technology at the heart of massive bureaucracies that deny any significance to anything that cannot be given a monetary value. From this relatively disembodied perspective, the global ecological crisis is seen as opportunities for further profits by creating new markets.[55] With the development of electronic media, including the internet, postmodernity has brought this dissociation of human cultural processes from ecological conditions to fruition, transforming much of 'culture' itself into commodities for consumption. As a consequence, most of humanity remains indifferent to the warnings of ecologists and climate scientists of the immanent destruction of the current regime of the global ecosystem on which civilization depends.

Democracy and the narrative of re-embodiments

The fusion of markets and bureaucracies has virtually destroyed the old opposition between the political left (who tended to promote bureaucracies to control, or even eliminate, markets) and the right (who tended to promote markets in place of or against bureaucracies). Those whose primary allegiance was to bureaucracies have joined forces with those whose primary allegiance was to markets, the consensual view now being that the common good is nothing more than growth of GDP, and liberty is nothing more than freedom to shop. Those whose primary allegiance was to democracy, betrayed by their former allies, are finding common ground with old opponents. The division that now matters is between those embracing a globalized economy fusing managerialism and market fundamentalism along with 'liberal individualism' – the individualism of irresponsible consumers – and those defending democracy. While the apologists for the globalization of corporations and the market have embraced the illusion of disembodiment as empowering and liberating and have found common ground with apologists for this around the world, those defending democracy are struggling to hold together the shreds of community. The struggle for democracy has become not only a struggle to re-empower communities but a struggle against the destruction of the natural and social environments of these communities. It is a struggle to regain control over communities' destinies against processes destroying the environmental conditions of their existence.

Despite the relatively powerless position of those promoting democracy, this realignment has overcome the confusion caused by past allegiances and facilitated a clearer vision of the democratic project. It has become apparent that at a fundamental level the struggle to protect and revive democracy, that is, to re-empower people as situated members of communities, is a struggle for re-embodiments. In India, this movement has been characterized by Vandana Shiva as 'earth democracy'; in China it has been characterized by Pan Yue as the quest for 'ecological civilization'.[56] Each of these developments provides evidence that we are moving into the Age of Re-Embodiments, re-embodiments likely to take a different form according to the unique situation and history of each community. It is necessary to consciously and actively work to overcome corporatocracy with a 'patchwork quilt' (to use Richard Norgaard's terminology)[57] in which people at local levels support a global system of decentralization and local empowerment. Their participants increasingly see local struggles as participation in this global struggle, and the lost plot of the struggle for democracy is being recovered, beginning to crystallize a new grand narrative. While the crisis of democracy has spawned new histories of democracy, most of these histories abstract from broader cultural history and do not mention the new field of environmental history.[58] What is needed is an account of the history of democracy in the broader context of nature. It is now possible to outline the framework that is emerging for such an integrated perspective. To begin with, it is necessary to be more specific about the meaning of democracy to show how this relates to culture more generally, and then show how the development of the culture of democracy should be understood in relation to the rest of nature.

Democracy requires history, philosophy and the arts to inspire people and enable them to coordinate their orientations and actions.[59] All complex actions involving many people are lived stories and require the telling and retelling the story of the action in which they are engaged, and then constraining action in accordance with the logic of these stories.[60] Stories are also central to the development of communities and institutions, and to maintaining their vitality. It is only through telling and retelling the history of communities and institutions that the point of their existence can be understood and revised. And stories are central to individuals in their efforts to orient themselves in a socially constituted world, to live life authentically and to refigure the stories they have inherited.[61] Democracy requires that these stories be open to question and revision by the public and be continually questioned and revised.

Through historical narratives the Ancient Greeks, the first people to explicitly defend democracy as such, examined the causes of conflicts, of failures and of greatness, while holding people responsible for their actions and orienting them to create the future, enabling them to build on the achievements of the past. This historical narrative included the achievements of philosophy, which provided a broader framework for the Greeks to orient themselves and a broader framework to defend democracy. By construing the cosmos as self-organizing and evolving through 'limiting the unlimited',[62] Anaximander, the first major Greek philosopher, challenged the received view that order has been created by a supposedly divine monarch.[63] Seeing people as part of nature, Anaximander held out the prospect of people governing themselves and taking responsibility for their own institutions, situating human history in the context of the cosmos.[64] The whole history of the development of democracy can be seen as unfolding the full implications of Anaximander's philosophy. His conception of the world has been rediscovered and developed with each effort to revive democracy and to overcome disembodiment fetishism, and in each case of such revival, the implications of Anaximander's vision have been realized more fully. Slaves, serfs, labourers and women have been emancipated from subjection as Anaximander foresaw. Moreover, the philosophies of Giordano Bruno at the end of the Renaissance, of Herder, F.W. Goethe and Friedrich Schelling at the end of the eighteenth and beginning of the nineteenth century and then of the process meta-physicians, C.S. Peirce, Henri Bergson, Aleksandr Bogdanov, A.N. Whitehead and John Dewey among others can all be seen as rediscoveries and further developments of this philosophy, in each case situating humans as historical agents within a creative nature and recognizing the freedom and significance of more and more people. Another revival of this way of thinking began during and after the Second World War, mainly within science but also in the philosophical biology and philosophical anthro-pology of philosophers such as Helmuth Plessner, Hans Jonas and Maurice Merleau-Ponty. With some setbacks, this has been gaining momentum ever since, challenging Cartesian dualism and creating a new alliance between science, the humanities and the arts. It is to a consideration of this new alliance that we now turn.

Re-embodiment and science

While this new alliance was proclaimed first in thermodynamics (by Ilya Prigogine and Isabelle Stengers), it is in ecology that it is coming to fruition. Ecology is now challenging the place accorded to physics as the ultimate reference defining science.[65] There are a number of elements being integrated in recent theoretical ecology: non-linear thermody-namics, complexity theory, hierarchy theory and biosemiotics, including eco-semiotics.[66] Such developments involve a new way of understanding the nature of life, and justify Lovelock's Gaia hypothesis, the claim that the earth itself is alive and has produced and is maintaining the conditions for life. These ideas have provided the basis for new developments in human ecology, characterizing humanity as a complex of processes, structures and semiosis (the production and interpretation of signs) within the global ecosystem. As Kalevi Kull argued, ecosystems are made of semiotic bonds, with more complex form of semiosis built upon, and presupposing, more primitive forms.[67] Cultural semiosis, associated with signs that are dissociated from immediate action, presupposes animal semiosis in which actions as interpretants of situations are signs, which in turn presupposes vegetative semiosis in which the interpretant is growth of form.[68] Human culture must then be seen as a development of what Jesper Hoffmeyer characterized as the 'semiosphere'.[69]

Morphogenesis, the generation of highly complex, ordered structure is a form of 'vegetative semiosis'. And as Christopher Alexander, an architect and complexity theorist, pointed out, 'the enormous and extensive co-adapative harmony of organisms in Nature is altogether due to morphogenesis'.[70] Such vegetative semiosis, which includes the human body, is presupposed by 'animal semiosis' or action, which in turn is presupposed by 'intellectual semiosis' or thought. It is an illusion to think that the meaning of abstract concepts could be understood except in relation to people's bodily engagement in the world and that it would be possible to capture the full richness of the world in abstract models. Indeed, it is a cultural disease to take these models as the source of reality. From this perspective, humanity is a very complex experiment by the global ecosystem characterized by a unique kind of semiosis whereby semiosis itself is reflected upon and interpreted. This second order semiosis enables humans to constitute their worlds as shared worlds in which individuals see themselves as components of the worlds of other subjects who could outlive them.[71] Defining oneself through proper recognition of others involves constraining of thought and action in taking into account the significance of others. This constraint is greatly augmented by the capacity to produce stories or narratives that allow people to develop complex forms of cooperation for projects over long durations, extending beyond the lives of individuals.

From the perspective of this philosophy, disembodiment is a denial of justice to people and nature and a blindness to the conditions for the existence of civilization. Disembodiment is a corruption of semiosis comparable to the semiotic corruption of cancerous tumours.[72] The alternative philosophy of re-embodiments projects a world of communities of communities at multiple scales with diverse human communities seen as participants in broader ecological communities, including Gaia.[73] The immediate task to realize this vision will be to redefine the end (telos) of society as being not the growth of GDP but the creation of vibrant communities that will cultivate authentic individuals and liberate them from the homogenizing, disembodying and destructive imperatives of the bureaucratized global market.[74] To this end the people need to be re-involved in discussing, formulating and evaluating policies, rediscovering, affirming and developing their own traditions – thereby avoiding the illusions created by 'experts' taking abstract models for reality and basing policy decisions upon these illusions that have been responsible for one social disaster and ecological failure after another.[75] It is necessary to re-embed humanity in nature so that people appreciate, adjust to and augment the dynamics and resilience of the ecosystems of which they are part. Re-embodying humanity will also involve reviving rural communities, fostering local self-sufficiency, reversing the balance of power between cities and the countryside and reversing the flow of people into the cities while fostering a high level of cultural development in rural communities. Brian Goodwin, the eminent theoretical biologist, summed up what is required in his last book, *Nature's Due.*

> The Great Work, the *Magnum Opus* in which we are now inexorably engaged, is a cultural transformation that will either carry us into a new age on earth or will result in our disappearance from the planet. The choice is in our hands. I am optimistic that we can go through the transition as an expression of the continually creative emergence of organic form that is the essence of the living process in which we participate. ... This Gaian Renaissance will lead to what Thomas Berry calls the Ecozoic Age, in which all inhabitants of the planet are governed by the principles of Earth Jurisprudence in an Earth Democracy.[76]

This cultural transformation will alter how people understand themselves at a visceral level, embodying this culture as part of their habitus. They will come to understand in the way they live and work, in the environments they build and in the way they organize themselves and participate in communities and organizations, in the way they interpret each situation and in the way they think, that they are embodied components of ecosystems, part of and responsible for the future of nature, participating in nature's semiosis and responsible for the resilience and vitality of their social and ecological communities. This is the grand narrative of the Age of Re-Embodiments.

Notes

1   S. George, *How the Other Half Dies: The Real Reason for World Hunger,* Harmonds-worth: Penguin, 1976. Re-released in 2009 with a postscript as a free online download.

2   K. Marx, *Grundrisse: Foundations of the Critique of Political Economy (Rough Draft),* trans. Martin Nicolaus, Harmondsworth: Penguin, 1973, p. 232.

3   J. Derrida, *Dissemination*, trans. Barbara Johnson, Chicago, IL: University of Chicago Press, 1981, p. 328.
4   R. M. Solow, 'The Economics of Resources or the Resources of Economics', *The American Economic Review*, 1974, 64 (2), 1–14, p. 11.
5   Z. Bauman, 'Time and Class', *Arena Journal*, New Series, 1988, 10, 69–84, p. 77.
6   Ibid.
7   S. Greenfield, *Mind Change: How digital technologies are leaving their mark on our brains*, London: Rider Books, 2015.
8   J. Baudrillard, *Symbolic Exchange and Death*, trans. I.H. Grant, London: Sage, 1993 [1976], p. 7.
9   See N. K. Hayles, *How We Became Posthuman: Virtual Bodies in Cybernetics, Literature, and Informatics*, Chicago, IL: University of Chicago Press, 1999.
10   J. Baudrillard, *Simulations*, trans. P. Foss, P. Patton and P. Beitchman, New York: Semiotext(e), 1983. In this Baudrillard suggests that there is no original reality to be simulated, but this should be seen as Baudrillard's strategy to outdo the postmodernists to highlight the absurdity of their views and make it impossible for them to dismiss him.
11   On the dissociation of economic models and the real world, see M. Perelman, *The Confiscation of American Prosperity*, Macmillan: Palgrave, 2007, Part 4.
12   The dynamics of this were revealed by Susan George in *How the Other Half Dies: The Real Reasons for World Hunger*, Harmondsworth: Penguin, 1977. For a more recent analysis, see V. Shiva, *Earth Democracy: Justice, Sustainability, and Peace*, Cambridge, MA: South End Press, 2005, esp. Ch. 1.
13   See J. Becker, *Dragon Rising: An Inside Look at China Today*, Washington D.C.: National Geographic, 2007, p. 141. See also C.K. Lee, *Against the Law: Labor Protests in China's Rustbelt and Sunbelt*, Berkeley, CA: University of California Press, 2007.
14   See World Bank, *Beyond Economic Growth*, Ch. 9 'The Growth of the Service Sector', 2000. Available online at: www.worldbank.org/depweb/beyond/beyond.htm (accessed 16th September 2015).
15   M. Wolf, 'Why it is so Hard to Keep the Financial Sector Caged', *The Financial Times*, 5 February 2008. Available online at: www.ft.com/cms/s/0/9987c5c4-d41f-11dc-a8c6-0000779fd2ac.html#axzz19TKMNVj4 (accessed 20th March 2013).
16   W.H. McNeill, *The Human Condition: An Ecological and Historical View*, Princeton, NJ: Princeton University Press, 1980, p. 6.
17   See H. and H.A. Frankfort, J.A. Wilson and T. Jacobsen, *Before Philosophy*, Harmondsworth: Penguin, 1964, p.118f.
18   J.A. Tainter, *The Collapse of Complex Societies*, Cambridge: Cambridge University Press, 1990, p. 175.
19   G. Roux, *Ancient Iraq*, 2nd ed. Harmondsworth: Penguin Books, 1980, p. 117ff.
20   W.H. McNeill, *Plagues and Peoples*, New York: Doubleday, 1989.
21   W.H. McNeill, *The Human Condition, An Ecological and Historical View*, Princeton, NJ: Princeton University Press, 1980.
22   On this see S.C. Chew, *The Recurring Dark Ages: Ecological Stress, Climate Changes, and System Transformation*, Lanham: AltaMira Press, 2007; R.J. McIntosh, J.A. Tainter and S.K. McIntosh (eds), *The Way the Wind Blows: Climate, History, and Human Action*, New York: Columbia University Press, 2000, and Tainter, op. cit.

23   T.F.H. Allen, J.A. Tainter and T.W. Hoekstra, *Supply-Side Sustainability*, New York: Columbia University Press, 2003, p. 138f.

24   See Tainter, op. cit., p. 49f.

25   Saint Augustine, *The City of God*, trans. M. Dodds, New York: Random House: 1950, Book 10, Ch. 14.

26   See A. Gare, *Nihilism Inc.*, Sydney: Eco-Logical Press, 1996, Ch. 3.

27   See T.F.H. Allen, J.A. Tainter and T.W. Hoekstra, op cit., p. 160.

28   R.B. Norgaard, *Development Betrayed: The End of Progress and a Coevolutionary Revisioning of the Future*, London: Routledge, 1994, Ch.11.

29   See E. Kamenka, *Bureaucracy: New Perspectives on the Past*, Oxford: Basil Blackwell, 1989, Ch. 1.

30   M. Bakhtin, *Rabelais and His World*, trans. H. Iswolsky, Bloomington, IN: Indiana University Press, 1984 [1965], p. 83.

31   On Pythagoras, see R. Rosen, 'The Church-Pythagoras Thesis', in *Essays on Life Itself*, New York: Columbia University Press, 1999, Ch. 4.

32   *Immanuel Kant's Critique of Pure Reason*, trans. Norman Kemp Smith, London: Macmillan, 1964, p. 20.

33   Aristotle, *Metaphysics*, 1015a 13–19, trans. R. Hope, Ann Arbor, MI: University of Michigan Press, 1960, p. 93.

34   Cited without original reference by F. Capra, *The Web of Life*, London: HarperCollins, 1996, p. 19.

35   S. Toulmin, *Cosmopolis: The Hidden Agenda of Modernity*, Chicago, IL: University of Chicago Press, 1992.

36   V.I. Lenin, *Collected Works*, Moscow: Progress Publishers, 1964, vol. XXVII, pp. 90–1.

37   See S.P. Huntington, *The Clash of Civilizations and the Remaking of World Order*, New York: Simon & Schuster, 2003, p. 51.

38   As noted by R.B. Norgaard, *Development Betrayed: The End of Progress and a Coevolutionary Future*, London: Routledge, 1994, p. 47. See also R.N. Adams, *The Eighth Day*, Austin, TX: University of Texas Press, 1988 and A. Hornborg, *The Power of the Machine*, Walnut Creek: AltaMira Press, 2001.

39   J.G. Herder, 'On the Cognition and Sensation of the Human Soul', in *Philosophical Writings*, Michael N. Forster (trans. and ed.), Cambridge: Cambridge University Presss, 2002, 187–243.

40   *Jefferson: Political Writings*, Joyce Appleby and Terence Ball (eds), Cambridge: Cambridge University Press, 1999, 206ff.

41   Kamenka, op. cit., p. 122.

42   Kamenka, op. cit., p. 125f.

43   See F.A. Hayek, *The Road to Serfdom*, Chicago, IL: University of Chicago Press, 1944, Ch. 5.

44   The classic study of this is Max Weber, 'Bureaucracy', in H.H. Gerth and C.W. Mills (eds), *From Max Weber*, London: Routledge & Kegan Paul, 1970, Ch. 8.

45   See J. Perkins, *Confessions of an Economic Hit Man*, New York: Plume, 2006, esp. p. 94.

46   Perkins provides an insider account of how this worked.

47   See D.C. Korten, *When Corporations Rule the World*, San Francisco, CA: Berrett-Koehler, 2001, second edition, and W.I. Robinson, *A Theory of Global Capitalism*,

Baltimore, MD: John Hopkins University Press, 2004. See also J. Perkins, *Hoodwinked*, New York: Broadway Books, 2009.

48   J. Galbraith, *The Predator State*, New York: Free Press, 2009.

49   S.S. Wolin, *Democracy Inc.: Managed Democracy and the Specter of Inverted Totalitarianism*, Princeton, NJ: Princeton University Press, 2010.

50   Shiva, op. cit., p. 152.

51   As Quentin Skinner showed in *Hobbes and Republican Liberty*, Cambridge: Cambridge University Press, 2008, esp. Chs 3 and 5.

52   See M.E. Hobart and Z.S. Schiffman, *Information Ages: Literacy, Numeracy, and the Computer Revolution*, Baltimore, MD: Johns Hopkins University Press, 1998.

53   B. Readings, *The University in Ruins*, Cambridge, MA: Harvard University Press, 1996.

54   R.H. Nelson, *Economics as Religion: from Samuelson to Chicago and Beyond*, University Park, Pennsylvania, PA: Pennsylvania University Press, 2001, p. xv.

55   See P. Mirowski, 'The Neoliberal Response to Global Warming', in *Never Let a Serious Crisis go to Waste: How Neoliberalism Survived the Financial Meltdown*, London: Verso, 2013, pp. 234–42.

56   See V. Shiva, *Earth Democracy: Justice, Sustainability, and Peace*, Cambridge, Mass.: South End Press, 2005, and Z. Wang, H. He and M. Fan, 'The Ecological Civilization Debate in China', *Monthly Review*, 2014, 66 (6), November, 37–59.

57   R.B. Norgaard, *Development Betrayed*, Ch. 14, 'A Coevolving Patchwork Quilt'.

58   See, for instance, J. Keane, *The Life and Death of Democracy*, New York: Norton, 2009, and D. Held, *Democracy and the Global Order*, Cambridge: Polity Press, 1995.

59   See A. Gare, 'The Arts and the Radical Enlightenment: Gaining Liberty to Save the Planet', *The Structurist*, nos. 47/48, 2007/2008, 20–7.

60   For the relationship between stories and action see D. Carr, *Time, Narrative, and History*, Bloomington, IN: Indiana University Press, 1991.

61   On stories or narratives, see A. Gare, 'The Primordial Role of Stories in Human Self-Creation', *Cosmos & History: The Journal of Natural and Social Philosophy*, 2007, 3 (1), 93–114.

62   See P. Seligman, *The Apeiron of Anaximander*, London: The Athlone Press, 1962, p. 58.

63   See C. Castoriadis, 'The Greek *Polis* and the Creation of Democracy', *Philosophy, Politics, Autonomy*, New York: Oxford University Press, 1992, Ch. 5.

64   J.-P. Vernant, *The Origin of Greek Thought*, Ithaca, NY: Cornell University Press, 1982.

65   R.E. Ulanowicz, *Ecology: The Ascendent Perspective*, New York: Columbia University Press, 1997, p. 6.

66   See S.N. Salthe, *Development and Evolution: Complexity and Change in Biology*, Cambridge, MA: MIT Press, 1993, and J.L. Lemke, 'Opening Up Closure: Semiotics Across Scales', *Closure: Emergent Organizations and their Dynamics*, Annals of the NYAS, New York: New York Academy of Science Press, 2000, pp. 100–11.

67   K. Kull, 'Ecosystems are Made of Semiotic Bonds: Consortia, Umwelten, Biophony and Ecological Codes', *Biosemiotics*, 2010, 3, pp. 347–57.

68   See K. Kull, 'Vegetative, Animal, and Cultural Semiosis: The semiotic threshold zones', *Cognitive Semiotics*, 2009, 4, Spring, pp. 8–27.

69   See J. Hoffmeyer, *Signs of Meaning in the Universe*, trans. Barbara J. Haveland, Bloomington, IN: Indiana University Press, 1993.

70   C. Alexander, 'Sustainability and Morphogenesis: The Rebirth of a Living World', *The Structurist*, 2007/2008, 47/48, pp. 12–19.

71   On the emergence of culture as a specific kind of semiosis, see W. Wheeler, *The Whole Creature: Complexity, Biosemiotics and the Evolution of Culture*, London: Lawrence & Wishart, 2006.

72   See A. Gare, 'The Semiotics of Global Warming', *Theory and Science*, 2007. Available online at: http://theoryandscience.icaap.org/content/vol9.2/Gare.html (accessed 11 March 2013).

73   See A. Gare, 'Toward an Ecological Civilization: The Science, Ethics and Politics of Eco-Poiesis', *Process Studies*, 2010, 39 (1), pp. 5–38.

74   See A. Vatn, *Institutions and the Environment*, Cheltenham: Edward Elgar, 2005. See also H.E. Daly and J.B. Cobb Jr., *For the Common Good*, Boston, MA: Beacon Press, 1994, second edition.

75   On this see F. Berkes, J. Colding and C. Folke (eds), *Navigating Social-Ecological Systems*, Cambridge: Cambridge University Press, 2003.

76   B. Goodwin, *Nature's Due: Healing Our Fragmented Culture*, Trowbridge: Floris Books, 2007, p. 177f.

Chapter 2

What is the age of re-embodiments?

Or, the victorious assertion of *loci standi* over the barbarism of *instrumenta movendi*

Ruth Thomas-Pellicer

From the living-organism/inert-matter binary pair into *Loci Standi*

Within ecological thinking, there is a strong – orthodox – and a weak – defiant – approach to the scientific habit of extricating 'living organisms' from 'inert matter'. The orthodox variant is utterly anthropomorphic. It presupposes a purposive stratification by which inert matter is presented as subservient to the needs of the more 'complex' creatures. In this ethos, a handbook of ecology is in the position to set out – and stir minimal controversy – that in the chemical cycle '[e]ach chemical element required for growth and reproduction must be made available to each organism at the right time, in the right amount, and in the right ratio relative to other elements'.[1] A similar outlook can be spotted in the mainstream approach to our ecocidal contemporaneity as captured in the official project of sustainable development. The epoch-making Brundtland Report repeatedly refers, along Cartesian–Newtonian lines, to an inert subservient backdrop called 'the environment'. 'Environment and development are not separate challenges; they are inexorably linked. Development cannot subsist upon a deteriorating environmental resource base; the environment cannot be protected when growth fails to take into account the costs of environmental destruction'.[2]

These initial *aporias* or impasses are aggravated when the scientific body remains utterly anthropocentric. Ecological economics too is anchored in the binary between living and inert, or biotic and abiotic, natures. This domain of knowledge has, on the one hand, achieved a major cognitive synthesis so as, rhetorically at least, to embed self-regulating markets in their larger ecological substratum. This theoretical achievement notwithstanding, ecological economics operates within a framework that still speaks the language of 'environmental resources and sinks' and 'natural resources and services',[3] responding thereby to the logic of the market not to that of the commons.[4] The anthropocentric and utilitarian resonances of all the scientific cases named hitherto are conspicuously evident: 'Such approaches are united in maintaining and propagating the idea of the "environment" [...] as essentially a mere setting for the human drama, most of which comprises a set of passive resources for the advancement of human interests, with the latter being the most, or even only, ethically considerable kind'.[5] It seems thus that the militant division between the living and the inert provides a metaphor for the 'natural world' by which allegedly higher forms of life operate against a supposedly passive receptacle, which they transform at will and to their complete advantage.

In this context it must be added that the scientific primacy lent to *logos, ratio*, renders human forms immediately superior to non-human ones. However, those strands of evolutionary theory that 'argue [...] that all sensible criteria of evolution must bear out the superiority of man'[6] must be read as deleterious spinoffs of the Western metaphysical trajectory. Evolution is not conterminous with superiority of more recent over former forms; it simply speaks of one of the multiple roads that generates diversity – another indeed being humanity's respect for this pulsating variety. Similarly, it must be added that '[i]t seems remarkably species-chauvinistic of psychologists to reserve for the human psyche qualities that are found everywhere outside it'.[7] Additionally, the Darwinian and Darwinistic premise of natural selection, according to which more gifted individuals are prone to draw more correct conclusions about the nature of the world, and thus are more capable of surviving and reproducing themselves in a way that they will become automatic models to the rest of the human species,[8] needs multiple qualifications. Geniuses of the highest calibre are often exposed to the most strident forms of ostracism, and only on occasions enjoy a following while alive, with their work often going utterly unrecognized if it challenges the interests of politico-economic cliques.[9] Selection does not rigidly stick to a logocentric pattern such as physicalism – 'the thesis that all things that exist in this world are bits of matter and structures aggregated out of bits of matter, all behaving in accordance with laws of physics'.[10] Rather, selection is the result of a number of contingent interactions salient among which are power relations.

The search for repeated patterns, the Western quest for a logocentric science derives from an urge to control and manipulate processes. This quest further construes evolution on the basis of an anthropocentric understanding of consciousness, rendering inferior those forms of embodiment that fail to articulate themselves logocentrically. Yet those non-logocentric forms, which include animals and indigenous people, appear to be the ones less prone to harm their neighbours. Schopenhauer[11] put it in a nutshell:

> The animal lives without any *Besonnenheit* [level-headedness]. [It] "remains constantly subjective, never becomes objective: everything that it embraces appears to exist in and of itself, and can therefore

never become an object of representation nor a problem for meditation. Its consciousness is thus wholly immanent. The consciousness of the savage man is similarly constituted in that his perceptions of things and of the world remain preponderantly subjective and immanent. He perceives things in the world, but not the world [—its objectification]; his own actions and passions, but not himself.

The visionary English world historian Arnold J. Toynbee considered that only in the context of a communion of saints the quandaries he identified as assailing his own contemporaneity – including ecocides – could be overcome. In a fairly evolutionary vein, the historian associates the ascent of the human species from its pre-humanoid form with the acquisition of consciousness and will. It was this critical breakthrough that, according to Toynbee, ushered in a critical 'intellectual and moral relativity [that renders e]ach of us [to] see [...] the Universe divided between himself (sic) and all the rest'.[12] Similarly, a good number of myths refer to the birth of the human condition as the moment when reason sets in and innocence is lost.[13] At any rate, the split between human ego and the rest of the universe, which has been exacerbated with the ascendancy of Western logocentrism, is evidently at the root of the global ecodebacle. In line with many authors in this volume, Toynbee compellingly emphasizes that such a life-destroying dualism is only amenable to rectification with the spiritual athleticism proper to the saint. It seems to me that non-logocentric entities – including animals and indigenous people – form, in a much more prominent manner than logocentric ones, the communion of spiritual athletes to which Toynbee vehemently appeals.

Be this as it may, let us resume our storyline on the orthodox versus defiant attitudes to the scientific habit of distinguishing between 'living organisms' and 'inert matter'. Somewhat surprisingly, Edward Goldsmith's *Ecological Way*[14] also constitutes a representative of the strong, orthodox, i.e. dualistic, approach. Goldsmith proposes to draw a dividing line between 'living things' and 'simpler things'. The former group, the ecological thinker goes on to contend, is 'capable of overcoming many of the constraints applying to the behaviours of'[15] the latter. One such limitation is the entropy law. In Goldsmith's analysis, the second law of thermodynamics should be excluded from the realm of ecology,[16] thereby preventing ecology from 'slavishly imitating the methodology of the physical sciences'.[17]

Gaia Theory, conversely, represents the weak approach. Lynn Margulis, intellectual mother of this strand of ecology,[18] argues that Gaia erases the animate versus inanimate boundary drawn in the 'natural' realm. This splitting is disparagingly assessed as a scientific convenience. Margulis sustains her argument by building on Aristotle:

Soil is not unalive. It is a mixture of broken rock, pollen fungal fila-
ments, ciliate cysts, bacterial spores, nematodes and other microscopic
animals and their parts. 'Nature', Aristotle observed, 'proceeds little by
little from things lifeless to animal life in such a way that it is impossible
to determine the exact line of demarcation'. Independence is a
political, not a scientific term.[19]

In living systems we have yet another instance where the distinction at stake
is effaced, namely Eigen's biochemist 'hypercycles'. Allegedly inert,
chemical hypercycles present the same pattern of self-organization and
originality as living systems. '[S]elf-replication – which is of course, well-
known for living organisms – may have occurred in chemical systems
before the emergence of life, before the formation of genetic structure
[...] The lesson to be learned here seems to be that the roots of life reach
down into the realm of nonliving matter'.[20] This suggests that the hierar-
chizing polarity of living organisms/inert matter is far too random to
warrant much more attention.

Partially endorsing the direction of Gaian thinking, and partially
endorsing a view that 'in the Age of the Industrial Revolution, human love
needs to be extended to include all components of the biosphere, inanimate
as well as animate',[21] the binary pair at hand emerges as so inaccurate as to
be irrelevant. At bottom, this division is exclusively grounded in the presence
of a chemical element, namely, carbon. The field of chemistry, to be sure,
establishes that life (on Earth) is carbon-based. The inanimate–animate
dualism is thus one immediately ruled by chemistry in the more extensive
context of modern science. This rigorous obedience to chemical composition entails overlooking alternative standpoints perhaps more significant in
light of our eagerness to enter a post-ecocidal age. The cleavage at hand
imperturbably discriminates entities that approached from distinctive
perspectives have important traits in common. For example, a rock and a
parrot may share the fact that both run non-ecocidal lives, while chemistry
classifies the two in opposed groups. The former is positioned in an animate
category while the latter in an inanimate one. From the perspective of our
planetary conjuncture it may indeed be that the classification ecocidal/non-
ecocidal is more relevant than the animate/inanimate one.

The radical philosopher of science Paul Feyerabend concluded that
realism is only tenable when we admit that 'quarks and gods [are] equally
real, but tied to different circumstances'.[22] Comparably, our planetary
contemporaneity begs that those with a concern for the state of the Planet
identify *non-ecocidal* entities and processes as real. In keeping with Gaia
Theory teachings, '[w]e can no longer think of rocks, animals, and plants
being separate. [...] There is a tight interlocking between the Planet's
living parts – plants, microorganisms, and animals – and its nonliving parts
– rocks, oceans, and the atmosphere'.[23] All these elements share their

capacity to provide secure stay for existence. They all conform to irreducible backbones that keep the whole of creation – perhaps magically – suspended in the cosmos. We should thus note that a scrupulous reading of the living–inert[24] binary pair has brought us to its transmutation into a multiplicity of secure stays. There are no entities that are alive as opposed to some that are dead.[25] Rather, there is a variety of life forms. Honouring their peculiar traits of 'necessity', 'irreducibility', 'security' and having recourse to classical Latin – as against English – as a putative *lingua franca* for a post-ecocidal age,[26] we shall designate these entities under the signifying phrase *locus standi,* in the singular, and *loci standi* in the plural. Plants, microorganisms, animals, rocks, oceans and the atmosphere have in common that they all are *loci*[27] *standi*[28] –'places of (secure) stay', even as they are always also animate, dynamic, changing.

A transmutation of the theist/atheist binary pair

There are many plausible avenues through which to characterize *loci standi.* The domain of ecology, to name one possibility, sets forth a taxonomy based on the different components of the Earth: the lithosphere (rocks and soils), the atmosphere, the hydrosphere (oceans, rivers, lakes, groundwaters and glaciers) and the biosphere (plants and animals).[29] Ecologists may alternatively concentrate on the tenets of distinct biomes and speak in terms of tropical rainforests, the tundra in Arctic regions and the evergreen trees in the coniferous forests. Here we are, on the other hand, deflecting 'the helpless enterprise of arriving at a philosophically (or, for that matter, scientifically) impeccable theory that will command the assent of all rational beings'.[30] Rather, our purpose here is to politicize a number of aspects of our contemporaneity so that the ecocidal tendencies deeply ingrained in Western/ized societies may be significantly reversed.

With a view towards politicizing certain (long-lost) dimensions of our present times, we shall initially characterize *loci standi* as those sites from which communities have ancestrally derived meaning. Victor M. Toledo's[31] description of vernacular peoples' cosmology sets the initial definition of *loci standi* into the apposite context:

> Indigenous people do not consider the land [a *locus standi*] as merely an economic resource. Under indigenous cosmovisions, nature [a pool of *loci standi*] is the primary source of life that nourishes, supports and teaches. Nature is therefore not only a productive source but also the centre of the universe, the core of culture and the origin of ethnic identity. At the heart of this deep bond is the perception [a *locus standi*] that all living and non-living things and natural [an abundance of *loci standi*] and social worlds [a plethora of *loci standi*] are intrinsically linked.

Toledo points to one of the cornerstones of *loci standi*. Throughout the millennia, communities have approached the latter with a profound sense of the sacred. This reverent attitude explains the reason why *loci standi* have recurrently performed their multifaceted roles as 'sites of secure stay'. Voluminous bodies of evidence in ecological anthropology and eco-ethnicity are conclusive in this respect.[32]

Social philosopher of religion Peter Berger[33] exposes, for his part, the reason why 'viewed historically, most of man's [sic] worlds have been sacred worlds: [...] it appears likely that only by way of the sacred was it possible for man to conceive of a cosmos in the first place'. The ecophilosopher Patrick Curry[34] further expounds on the reasons for this historical constant. The immense variety of *loci standi* can 'never be exhausted, finally or comprehensively explained, or completely mastered. This entails an ultimate mystery which constitutes the source and terminus of all meaning and value'. On this count, the cultural functions of *loci standi* largely overlap with those of *mythoi*. Philosopher of religion Raimon Panikkar[35] attributes to myth a plural, original and ultimate source of meaning: 'We live on myths' shoulders'. Just as we characterise *locus standi* as an 'irreducible place of secure stay', so too Panikkar portrays myth as the enactment of the original and primal – albeit not necessarily foundational. From this exposition it follows that myth does not lend to seizure, let alone manipulation. 'The transparency of myth consists in the fact that we see through a myth, but we cannot see it'.[36]

The irreducibility of *loci standi* and *mythoi* exempts both from the need to be proven – neither of them are objects of either demonstration or proof. Rather, both inexorably exact overt acknowledgement and inviolable respect. Inextricably plural, neither *loci standi* nor *mythoi* are amenable to their reduction to a common denominator. Each generates a constellation without any stable centre or hierarchy. *Loci standi* and *mythoi* are irrevocably incommensurable with, respectively, each other. In this context, Fred Block's[37] vocal indictment of the secular character of modern societies is wholly vindicated – the latter no longer exhibit the ancestral reverence for the entities that we have come to name *loci standi*:

> It is simply wrong to treat nature and human beings as objects whose price will be determined entirely by the market. Such a concept violates the principle that has governed societies for centuries: nature and human life have almost always been recognized as having a sacred dimension. It is impossible to reconcile this sacred dimension with the subordination of labour and nature to the market.

In this ethos an urgent call to irreversibly commit a theocide of the objectifying God (*qua* the metaphysical device that reverts multiplicity back to One and becoming to Being) must be advanced. The mere adoption of an atheist attitude shall prove insufficient for our purposes here. The simple denial of a monotheist figure, without further structural decentrings, fails to contest the ontologically objective configurations that God's presence, *qua* ultimate foundation of reality, keeps steadfastly in place. Atheism irremediably operates in theism's shadow – it is its mirror. 'When [...] God appears as unnecessary for modern science, atheism is around the corner [...] the world is ruled by laws where God does not intervene, except at the beginning'.[38] Once we dare analyse the binary pair theism/atheism we automatically grow aware of the untenability of the ontological edifice sustained with God's objectivist metaphysics. For 'what is put into question is precisely the quest for a rightful beginning, an absolute point of departure, a principal responsibility'.[39] This suspicious probing leads us to a theological decapitation. This murderous act releases, in turn, the historical – situated – and analytical space. It lays bare the concrete individuals, sources and processes that are concealed behind this overarching objectivism. No wonder that '[o]nce the local character of all philosophical authority and subordination is exposed [...] the ultimate or unconditional nature of authority [...], starting with *logos* and truth, is removed. All such authorities have to be reconfigured locally, historically, practically, including politically but also psychoanalytically [...] All authorities then become local authorities'.[40] It is in this decompressed – as opposed to repressed – space that we are free again to appreciate an inextricable plurality and originality. As a result, we are free once more to engage in animist worshipping practices. '[T]he end of [Western] metaphysics and the death of God [is] above all the rebirth of the sacred in its many forms'.[41] In the shambles of Western metaphysics, we find the places of secure stay shining in a plurality of sanctuaries. 'When the origin has revealed its insignificance, as Nietzsche says, then we become open to the meaning and richness of proximity',[42] namely, to the plural *loci standi*. As Nietzsche further claims, it is the religious spirit that has forcefully beheaded the metaphysical God. 'The death of God [...] is an effect of religiosity'.[43]

The multifaceted intellectual Gregory Bateson speaks of the ecology of mind by reversing the 'traditional God-down hierarchy of being',[44] which operates as the transcendent divine mind that bestows identity on the natural world. In step with the panpsychist tradition proper of the likes of Teilhard de Chardin, Bateson claims that mind, chiefly in the form of 'perception and awareness'[45] is present in 'all the processes that produce, for instance, healing in organs, growth in organisms, development in societies, or balance in large ecosystems'.[46] Yet Bateson is relevant for our purposes here in so far as he undertakes a metaphysically immanent U-turn. He disregards Being and speaks of a multiplicity that far from conforming to static attributes is dynamically defined. In many senses, Bateson's mental processes incarnated in palpable physical forms prefigure *loci standi*. Like Bateson's mental processes, *loci standi* gain their identity autonomously – liberated from a metaphysical objective figure. Batesons' emphasis on the mental or the 'processes of informational interaction *between* components'[47] may be seen as one aspect of the securing-stay feature characteristic of loci *standi*.

Some examples of *loci standi* are the genome; local species; non-GM seeds; macronutrients such as nitrogen, phosphorous and water; ecosystems; the hydraulic cycle; the Earth's carbon-cycling capacity; low entropy; customary law, and local and bioregional politics; animist worldviews are also a form of *loci standi*; so is any philosophy or worldview that does away with provincialism, while preserving that which is authochthonous.

Loci standi and their *naturae*

Loci standi enjoy the self-direction for self-maintenance, development and self-replication.[48] In a bicycle, that is, a non-*locus standi*, 'the link between pattern and structure is in the mind of the designer'.[49] Contrastingly, in a *locus standi* 'the pattern of organization is always embodied in the organism's structure, and the link between pattern and structure lies in the process of continual embodiment'.[50] With Murray Bookchin[51] we can assert that the cosmic drama, made up of a great multiplicity of *loci standi*, 'is "directed" not by a deity exogenous to it or by a divine "architect" who fashions it; rather, it would be a *self*-directed and self-*unfolding* drama, whose "finality" is as much an inherent property of substance as is motion, [altogether] yield[ing] to the world particulars [i.e., *loci standi*] in their wholeness and fullness as a rich unity of diversity'. Yet *loci standi* are neither origin (*arch*) nor end (*telos*), rather, they are above all original, creative, inventive places of stay. While myths 'speak of and to the original',[52] *loci standi* embody, embed and situate the original. They sustain the cosmos by rendering it perennially singular. Their distinctive ability is to play out their gifts with creativity and originality.[53] We shall say that these are elements endowed with a *natura*[54] all their own or a capacity to willingly come into being, create, preserve, destroy and transform themselves. *Loci standi* unrestrictedly cut across the Cartesian monolithic *res extensa*, purportedly made up of logocentrically realist essences. *Qua* primal sources, *loci standi* recurrently transpose – 'go back to'[55] – a spatiality that is not so much a spacetime experience as a qualitative – and quantitative – status: they form 'a realm of wild, inimitable, and [in principle] inviolable reality'.[56]

Each *locus standi* transfigures abstract – read: disembodied and disembodying – notions of space and unidirectional time into irreducible and repetitive – albeit never quite identical – shapes and rhythms of secure stay. In this respect, we can draw a telling simile between *loci standi* and the artisans in Buddhist traditions who, generation after generation – conveying a sense of timelessness – are in charge of reproducing the image of the Buddha. Like these committed workers, *loci standi* are 'steps in a chain that resists [unidirectional] time through which sacred forms manifest themselves'.[57] The irreducibility of *loci standi* rather re-reads unidirectional cosmic time as the pace proper to the cosmos *qua* a *locus standi* all its own.

Logocentrism turns *natura* into a passive, feminine monolith crisscrossed with repetitive and universal laws.[58] *Loci standi*, by contrast, embody, embed and contextualize *natura*, releasing thereby their intrinsic *naturae*.

A transmutation of the anthropocentrist/ecocentrist binary pair

Loci standi not only subvert the living organisms/inert matter splitting and the theistic/atheistic antagonism. Additionally, these are strategic transmutators of the anthropocentric/ecocentric binary pair. In two passages of his *Ecological Ethics*, Curry manifests the cognitive (and, for that matter, axic, ethical) *aporias* that holding to the divide entails. One is a reflection on the part of the author, while the other comes in the form of a few remarks on the ecocentric ethical body of J. Baird Callicott. In having laid out the tenets of ecocentrism, Curry[59] prudently cautions us against the fact that 'an appeal to anthropocentric value (human self-interest) may be an unavoidable part of the argument for an ecocentric outcome. It follows that an ecocentric ethic alone will probably not suffice to save an Earth fit for life as we know it; but that also won't happen without one'. Subsequently, Curry approvingly quotes Callicott's reflection on the vast dimensions of our embodied condition. Interestingly, these appear hard to theorize precisely given the Cartesian dividing line meticulously traced between a human realm and a natural one: 'As Callicott has rightly pointed out, any distinction between "inner" and "outer" or "self" and "other" is strictly relative and never ultimate, except as a modernist fantasy: "it is impossible to find a clear demarcation between oneself and one's environment... The world is, indeed, one's extended body"'.[60]

To be sure, *loci standi* bring to a new horizon Curry's problematized partition of anthropocentric *contra* ecocentric value, Callicott's tentative characterization of our embodied condition – to which Curry subscribes, and the utilitarianism prevalent in ecological economics mentioned several paragraphs above. *Loci standi* are neither ecocentric nor anthropocentric entities; rather, they are polycentric: they denote multiple foci. 'Oneself' (an anthropocentric entity), 'one's environment' (in the quote above meant as an ecocentric entity)[61] and 'environmental resources and sinks' are 'irreducible places of secure stay' or 'sites of firm repose'. Humans, animals, plants, minerals, when not stopping other species from thriving, all appear as places of firm stay. So do the Earth and the universe. They all conform to *loci standi*, which relate to one another in a network.

The advent of metallurgy and the historical necessity to register *instrumenta movendi*

As noted above, Goldsmith positioned himself against the validity of the entropy law in ecology. This is a polemical position to say the least. Arthur Eddington noted that 'if your theory is found to be against the second law of thermodynamics I can give you no hope; there is nothing for it but the collapse in deepest humiliation'.[62] Conceivably, though, there is no need to engage in this controversy. Our scientific evasion may be compensated for with the following observation. One of the arguments on which Goldsmith sustains his resistance to the dominance of entropy lies in the fact that '[l]ife probably began on this planet three thousand million years ago and since then [...] it has developed in complexity, diversity and stability'.[63] This tendency towards more complex life-forms, Goldsmith goes on to argue, disconfirms the appearance of randomness – so profoundly characteristic of the entropic process[64] – and with it the existence of the entropy law in 'nature'.

It is true that Goldsmith's anti-entropic argument fails to disprove the second law of thermodynamics. Indeed, the pioneering ecological economist Georgescu-Roegen regards the entropy law as 'the only clear example of an evolutionary law'.[65] On the other hand, Goldsmith's account is of great pertinence for our purposes here. His sweeping statement regarding the sustained development of diversity and complexity is largely accurate. In the lifespan implied by Goldsmith, either human forms of life were nomadic or most communities were tilling the soil *with only moderate levels of artifactual complexity* while being informed by relatively localized sources of information.

Goldsmith's position lends to controversy from the moment when artifacts gain in complexity and the sources of information cease to be exclusively local. Historically, this is known as the Neolithic Revolution and *a fortiori* as the practice of metallurgy.[66] The Neolithic Revolution is the moment when sufficient surplus is created and the division of labour as well as centralized forms of control are technically possible. As Toynbee states, '[i]n a society which produces an economic surplus, a division of labour is possible. Specialist minorities, freed from the work of food production, can monopolize social tasks which were formerly the responsibility of all participants in a society'.[67] Yet it appears that '[p]rimitive agriculture had produced no surplus food and therefore no reserves for maintaining specialists; the only division of labour had been between men and women, and each local community had been self-sufficient'.[68] While technology was already apparent in the Upper Palaeolithic, progressively, the development of technology marks a relevant cultural trend that comes to characterize the ecocidal trajectory of the Western/ized world. '[I]mprovements in tools of all kinds have accelerated, and, though there have been local and temporary pauses, and even relapses, acceleration has been the paramount tendency in the history of technology'.[69]

Among the abundance of sophistications and turning points in the technological exponential course, a number of authors (Toynbee, Derrida, Georgescu-Roegen) emphasize the advent of metallurgy. It is metallurgy, 'a full-time occupation',[70] that spawns all sorts of artifacts. 'Metallurgical lore is the first approximation to international science, and metallurgy destroys Neolithic self-sufficiency – requiring, as it does, not only smiths but miners, smelters, and carriers'.[71] Jacques Derrida, for his part, associates the constitution of society with the metallurgical penetration of the Earth: '[T]he violence that takes us toward the entrails of the earth, the moment of mine-blindness, that is, of metallurgy, is the origin of society [and] of the supplement'.[72] With the birth of civilization, *loci standi* are steadily replaced with their supplements. Culture 'destroys very quickly the forces that Nature has slowly constituted and accumulated'.[73] This first offshoot of the practice of metallurgy indicates that there are two bases on which to build a community: A non-ecocidal one and an ecocidal one. Each is largely determined by the intensity of technology given the multiple implications of the latter on the Planet's embodied condition.

The dawn of metallurgy, to be sure, represents a fierce assault on *loci standi*. 'In agriculture and animal husbandry, Man's technological power and Nature's productivity are in equilibrium. With the invention of metallurgy, Man's technological ability began to make a demand on Nature [*loci standi*] that she is [they are] not capable of satisfying throughout the time for which the biosphere can still continue to be habitable'.[74] Toynbee indeed retells that the practice of metallurgy represents a major disruption of *loci standi*. His retrospective admonition is vocal in this respect: 'If we think of the last 10,000 years of human history in terms of the 2,000 million years of mankind's potential expectation of life, we may perhaps conclude that it would have been better for our descendants if metallurgy had never been invented, and of Man, after having attained the Neolithic level of technology, had not succeeded in raising himself higher in terms of technological achievement'.[75]

Toynbee's concerns are in line with the findings of the role played by the entropy law in the economic process. '[T]he fact remains that the higher the degree of economic development, the greater must be the annual depletion [of the stock of terrestrial low entropy] and, hence, the shorter becomes the expected life of the human species'.[76] Since the dawn of metallurgy, along with *loci standi*, a second category of knowledge turns irreducible in our post-ecocidal toolkit. Or, phrased in Bruno Latour's words, we would prove poor anthropologists if we failed to register this second category: 'Those who are incapable of explaining the irruption of objects into the human collective, along with all the manipulations and practices that objects require, are not anthropologists, for what has constituted the most fundamental aspect of our culture, since Boyle's day, eludes them'.[77] On the basis of our account, we shall extend Latour's statement back to the day of metallurgy.

The objects at issue are tools that operate to the detriment of *loci standi*. They are also forms of control that uproot ecoregionally embedded governance, and cognitive processes that are foreign to contextualized-to-an-ecoregion worldviews. These artifacts now described seem to have one element in common. Namely, they all conform to tools that offer a kind of mobility that is both additional and supplementary to that provided by *loci standi* but at the high price of destabilizing the latter. We may also refer to these tools of mobility by way of their Latin equivalent, namely, as *instrumentum movendi* in the singular and *instrumenta movendi* in the plural.

We shall at this stage note that the irreducibility of *instrumenta movendi* prominently determines our inquiry into history. It seems that, since the advent of the Neolithic Revolution and metallurgy, a world history responsive to our embodied condition should be broadly guided by the following queries: under what cognitive frameworks, through what collective – both public and private – institutions, what actors provoke and justify what processes of dis-/re-embodiments? Or, to turn the question around, by means of what *loci standi* and *instrumenta movendi* are communities agreeing to live? The greater the amount of *instrumenta movendi* in circulation, the more processes of disembodiment underway, and the fewer *loci standi* left to secure stay. The suggested set of analytical queries neatly comes to capture the existential challenge that the West and the world at large – given the outspread influence of the former – faces. This cross-roads can be further characterized via Nietzsche's provocative hypothesis of eternal recurrence, according to which one's choices return in distinct ways: '"Are you living your life in such a way that you would will to live everything again the same way for all eternity?"'[78] Namely, have we truly approved to live with such an uncontrolled proliferation of *instrumenta movendi*? Shall in the future neither *instrumenta movendi* nor *loci standi* be left in view of the dependence of the former on the latter?

Instrumenta movendi should not be exclusively associated with technological artifacts. Rather, this rubric is much more extensive. It encompasses both objects and actions that destabilize *loci standi*. Some examples of *instrumenta movendi* are automobiles such as cars and trucks; centralized – in contradistinction with federal and *a fortiori* bioregional – forms of governance, such as state governments, central banks, the European Union, the World Bank and the World Trade Organization; ideologies with universalist intentions such as liberalism, Marxism, Islam and Christianity;[79] all concepts, idealizations defining Western metaphysics such as subjective idealism[80] and the binary pairs endemic to the former – including sustainable development/productivity and culture/nature.[81]

The victorious assertion of *loci standi* over the barbarism of *instrumenta movendi* or our unequivocal entrance to the age of re-embodiments

Polanyian economic anthropology[82] retells that the self-regulating market economy of the nineteenth century emerged as a *novum* in human history. Societies ruled by self-regulating markets constitute social organizations that lend priority to productive processes and thus exhibit prominent levels of disembodiment. These societies systematically feed on tools of mobility or *instrumenta movendi*. Or, to phrase it reversely, when a society exhibits heightened levels of disembodiment, we shall say that it is *instrumenta-movendi* oriented and that it runs on the mode of production. Correlatively, when a *cultus* exhibits full embodiment we shall say that it is *loci-standi* oriented and that it runs on the mode of re-production. Against Goldsmith, but with Georgescu-Roegen and other ecological economists,[83] one must contend that in *instrumenta-movendi*-oriented settlements the entropy law turns into a visibly restrictive parameter to be carefully taken into account. The ecological economist Tim Jackson puts it in a nutshell: 'the second law of thermodynamics is [...] a startling and important piece of physics for the industrial economy'.[84] Jackson further accounts for the reason why the dissipative effects of this law apply to *instrumenta-movendi*-oriented settlements, but appear dormant in *loci-standi*-oriented ones.[85] The 'self-regulat[ing]' property common to the entropic process 'is being threatened, in many cases, by the impacts of the industrial economy'.[86]

There is, nonetheless, a parallel reading to the last statement. Namely, the laws of thermodynamics are rooted in the mode of production – as opposed to that of re-production. 'Thermodynamics sprang from a memoir of Sadi Carnot in 1824 on the efficiency of steam engines'.[87] The birthplace and whole *raison d'être* of thermodynamics orbits around industrial productive processes. Not surprisingly, it is meta-industrial work, the ecofeminist Ariel Salleh incisively avers, that 'whether domestic care or organic farming [...] generates a vernacular replicating and reciprocating the thermodynamic circuits of nature'.[88] Industrialization represents a phenomenon only to be put on a par with the Neolithic Revolution at the technological level, and with metallurgy at the larger cultural one. 'The Neolithic technological revolution [...] was a technological change of the same order of magnitude and momentousness as the modern Western Industrial Revolution [...] There was no comparable technological change during the intervening age'.[89]

The entropy law is instructive in a second sense. In view of the constraints that it imposes on societies ruled by the mode of production, 'any use of the natural resources [*loci standi*] for the satisfaction of nonvital needs means a smaller quantity of life in the future. If we understand well the problem, the best use of our iron resources is to produce plows or harrows as they are needed, not Rolls Royces, not even agricultural tractors'.[90] In other words, on the one hand, the productive factory provides the wholesale artillery to formulate the entropy law; and, on the other, it jeopardizes a kind of 'living structure' that is not related to those 'living organisms' that gain their meaning out of the living organism-inert matter dyad proper to ecology's Cartesian nature. Rather, disembodying societies that run on the mode of production represent a threat to *loci standi*. This is one of the capital messages advanced by thermodynamics. The increasing number of menaces posed to *loci standi* and their actual disappearance have taken to bits the Cartesian metaphor, according to which the physical world is the inert backdrop to an irrestrictively productive factory. There is no inert backdrop, but multiple, necessary and irreducible *loci standi* that are altered and eventually annihilated as *instrumenta movendi* are deployed in varyingly amounts.

A swift entrance to the Age of Re-Embodiments, which seems to be a pressing demand not on future but *current* generations, entails the breaking down of large settlements into *loci-standi*-reliant communities and a substantial reduction of the barbarian *instrumenta movendi* now in circulation. 'By binding more and more of life [and giving birth to *instrumenta movendi*] in a form in which it cannot reproduce life [*loci standi*], capitalism, and a complicit modernity, disturbs the ecological balance. How this balance is righted remains to be determined. But there are now few on the planet who dispute that the balance needs to be corrected in this beleaguered present'.[91] The concerted and prompt building of the Age of Re-Embodiments as the precursor to a wholesale access to an Ecozoic Era[92] may be the Planet's only realistic hope.

Notes

1   D. B. Botkin and E. A. Keller, *Environmental Science: Earth as a Living Planet*, Hoboken, NJ: John Wiley and Sons, 2005; 5th edition, p. 100.

2   World Commission on Environment and Development (WCED), *Our Common Future*, Oxford: Oxford University Press, 1987, p. 37.

3   J. Martinez-Alier, *The Environmentalism of the Poor: A Study of Ecological Conflicts and Valuation*, Cheltenham, UK and Northampton, MA, US: Edward Elgar, 2002, pp. 44, 70, 216.

4   I. Illich, 'Silence is a Commons', Asahi Symposium Science and Man – The computer-managed society, Tokyo, Japan, 21 March 1982. Available online at: www.preservenet.com/theory/Illich/Silence.html (accessed 30 April 2015).

5   P. Curry, 'Re-Thinking Nature: Towards and Ecopluralism', *Environmental Values*, 12: 3, 2003: 337–60, p. 338.

6   T. Dobzhansky quoted in N. Georgescu-Roegen, *The Entropy Law and the Economic Process*, Cambridge, MA: Harvard University Press, 1974 [1971], p. 128, n35.

7   P. Curry, *Defending Middle-Earth: Tolkien, Myth and Modernity*, New York: Houghton Mifflin Company, 2004, pp. 126–7.

8   S. Hawking and L. Mlodinow, *Brevíssima història del temps*, trans. by David Jou, Barcelona: Columna Edicions, 2005, p. 23.

9   Wolfgang Amadeus Mozart died at a young age in dire poverty. Antoni Gaudí-Cornet did not design a single building in his birth town, Reus (Catalonia, Spain), and left no descendants.

10   J. Kim, 'The Mind-Body Problem at Century's Turn' in B. Leiter (ed.), *The Future for Philosophy*, Oxford and New York: Oxford University Press, 2006, pp. 129–52, p. 129.

11   Quoted in R. M. Bucke, *Cosmic Consciousness: A Study in the Evolution of the Human Mind*, Milton Keynes: Book Jungle, reprinted [1905], p. 20.

12   A. J. Toynbee, *A Study of History*, London: Oxford University Press and Thames and Hudson, 1977 [1972], p. 87.

13   Cf. R. Panikkar, *Mite, Símbol, Culte*, Opera Omnia Raimon Panikkar, vol IX, tom I, Barcelona: Fragmenta Editorial 2009, p. 81.

14   E. Goldsmith, *The Way: An Ecological World-View*, Athens, GA: University of Georgia Press, 1998; revised and enlarged edition, p. 443.

15   Ibid.

16   Ibid., Appendix One.

17   Ibid., p. 144.

18   Along with James Lovelock – its intellectual father.

19   Quoted in M. Midgley, *Gaia: the Next Big Idea*, London: Demos, 2001, p. 17.

20   F. Capra, *The Web of Life: A New Scientific Understanding of Living Systems*, New York: Anchor Books, 1996, p. 94.

21   A. J. Toynbee, *Mankind and Mother Earth: A Narrative History of the World*, Oxford: Oxford University Press, 1976, p. 594.

22   P. Feyerabend, *Farewell to Reason*, London and New York: Verso, 1987, p. 125.

23   Capra, op. cit., p. 104.

24   'Inert' and 'non-living' are used in this essay somewhat indistinctively, although 'inert' is preferred following the etymological rendering of the word. 'Inert' signifies 'without "ars"'or 'skill'. Synonyms of inert are thus 'unskilled', 'inactive', 'helpless', 'sluggish', 'worthless'. To be sure, the entities that the Western tradition downgrades and opposes to the living ones are those that are conceived as 'unproductively worthless'.

25   Here 'dead' is used as synonym with 'inert'.

26   I admit that my Catalan background has some bearing on the proposal to turn classical Latin into a post-ecocidal *lingua franca*.

27   The meanings of the noun '*locus, loci*' relevant to us are 'seat, rank, position; place, territory, locality, neighbourhood, region; position, point, site; part of the body; region, places (pl.); places connected with each other.'

28   '*Sto*' the infinitive form of '*standi*' means 'to stand, to stand still, to stand firm; to remain, to rest'. '*Standi*' is the future passive participle form of '*sto*' in genitive form. A literal translation would be 'will have stood firm'. In a language like English that no longer inflects either nouns, adjectives or pronouns through their different cases, one rather says 'of firm stay'.

29   Botkin and Keller, op. cit., pp. 74–8.

30   P. Curry, *Ecological Ethics: An Introduction*, Cambridge, UK and Malden, MA: Polity, 2006, p. 67.

31   V. M. Toledo, 'Indigenous Peoples and Biodiversity', paper presented at the Congrés de Biodiversitat, Institut d'Estudis Andorrans, 1999.

32   Ibid.; Goldsmith, op. cit., ch. 63. In this context it is easy to understand why M. Foucault (*The Order of Things: An Archaeology of the Human Sciences*, London: Tavistock Publications, 1980 [1970], pp. 373–87), J. Derrida ('Structure, Sign and Play in the Discourse of the Human Sciences', in *Writing and Difference*, trans., with an Introduction and Additional Notes, by Alan Bass, Chicago, IL: University of Chicago Press, 1978, pp. 278–93, p. 282.) and B. Latour (*We Have Never Been Modern*, trans. by Catherine Porter, Cambridge, MA: Harvard University Press, 1993), all claim that ethnology, the science that studies non-Western communities, is in the position of outsider with respect to the remaining human (and natural) sciences. '[E]thnology situates itself in the dimension of the historicity [...] which is the reason why the human sciences are always being contested, from without, by their own history' (Foucault, 1980 [1970], op. cit., p. 376). Ethnology is an incoming source of non-Western assumptions, which destabilize the certainty with which modern science is portrayed.

33   P. L. Berger, *The Sacred Canopy. Elements of a Sociological Theory of Religion*, New York: Anchor Books, 1990 [1967], p. 279.

34   P. Curry, 'Post-Secular Nature: Principles and Politics', *Worldviews: Environment, Culture, Religion*, 11 (2007), pp. 284–304, p. 284.

35   Panikkar, op. cit., p. 115, own translation from the Catalan rendering.

36   Ibid., p. 117.

37   F. Block, 'Introduction' in K. Polanyi, *The Great Transformation: The Political and Economic Origins of Our Time*, Boston, MA: Beacon Press, 2001 [1944], pp. XXV, XXVI.

38   Panikkar, op. cit., p. 99, own translation from the Catalan rendering.

39   J. Derrida, 'Différance', in *Margins of Philosophy*, trans., with Additional Notes, by Alan Bass, Chicago, IL: University of Chicago Press, 1984a [1968], pp. 1–27, p. 6.

40   A. Plotnitsky, *Reconfigurations: Critical Theory and General Economy*, Gainsville, FL: University Press of Florida, 1993, p. 153.

41   G. Vattimo, *After Christianity*, New York: Columbia University Press, 2002; translated by Luca d'Isanto, pp. 21, 23.

42   G. Vattimo, *The End of Modernity*, Baltimore, MD: John Hopkins University Press, 1991 [1988], p. 177.

43   Vattimo, *After Christianity*, op. cit., p. 26.

44   N. G. Charlton, *Understanding Gregory Bateson: Mind, Beauty, and the Sacred Earth*, Albany, NY: State University of New York, 2008, p. 35.

45   Ibid., p. 32.

46   Ibid.

47   Ibid.

48   M. Faber, R. Manstetten and J. Proops, 'On the Conceptual Foundations of Ecological Economics: A Teleological Approach', in M. Faber, R. Manstetten and J. Proops, *Ecological Economics: Concepts and Methods*, Cheltenham, UK and Northampton, MA, USA: Edward Elgar, 1998 [1996]): 168–88, p. 175; B. Allen, *Knowledge and Civilization*, Cambridge, MA: Westview, 2004, p. 63.

49   Capra, op. cit., p. 160.

50   Ibid.

51 M. Bookchin, 'Sociobiology or Social Ecology', in *Which Way for the Ecology Movement?*, Edinburgh and San Francisco, AK Press, 1994, pp. 49–75, pp. 62–3.

52 J. Caputi, 'On the Lap of Necessity: A Mythic Reading of Teresa Brennan's Energetics Philosophy', *Hypatia*, 16:2, Spring 2001, p. 5.

53 I. Prigogine and I. Stengers, *Order Out of Chaos: Man's New Dialogue with Nature*, Shambhala/Boulder and London: New Science Library, 1984.

54 *Natura, nascor, nasci, natus sum*, 'to be produced spontaneously, to come into existence/being, to spring forth, to grow, to live, to be born, begotten, formed, destined'. Available online at: http://thesaurus.babylon.com/natos?&tl= (accessed 30 April 2015).

55 Caputi, op. cit., p. 4.

56 Ibid.

57 A. Shearer, *Buda: Un corazón inteligente*, trans. from the original English into Spanish by Pablo Badía Mas, Madrid: Debate, 1993, p. 88, my own translation back to English.

58 A. Salleh, *Ecofeminism as Politics: Nature, Marx and the Postmodern*, London and New York: Zed Books Ltd, 1997; A. Salleh, 'Embodied Materialism in Action: An Interview with Ariel Salleh', in G. Canavan, L. Klarr and T. Vu (eds), *Polygraph: Special Issue on Ecology and Ideology*, 2010, Fall, No 22.

59 P. Curry, *Ecological Ethics: An Introduction*, Cambridge, UK and Malden, MA: Polity, 2006, p. 46.

60 Ibid., p. 66.

61 However, one's environment may equally be an 'anthropocentric' entity. It all depends on whether the environment at issue is filled with *instrumenta movendi* or *loci stundi*.

62 Quoted in M. Faber, R. Manstetten and J. Proops with S. Baumgärtner, 'Entropy: A Unifying Concept for Ecological Economics', in M. Faber, R. Manstetten and J. Proops, *Ecological Economics: Concepts and Methods*, Cheltenham, UK and Northampton, MA, USA: Edward Elgar, 1998 [1996]), pp. 95–114, p. 95.

63 Goldsmith, op. cit., p. 440.

64 Prigogine and Stengers, op. cit.

65 Georgescu-Roegen, op. cit., p. 128.

66 Toynbee, *A Study*, op. cit., p. 44–51.

67 Ibid., p. 44.

68 Ibid., p. 48.

69 Toynbee, *Mankind*, op. cit., p. 39.

70 Ibid.

71 Ibid.

72 J. Derrida, *Of Grammatology*, trans. by G. C. Spivak, Baltimore, MD: The Johns Hopkins University Press, 1997, p. 149.

73 Ibid., p. 151.

74 Toynbee, *A Study*, op. cit., p. 44.

75 Toynbee, *Mankind*, op. cit., p. 44.

76 N. Georgescu-Roegen, 'The Entropy Law and the Economic Problem', in H. E. Daly and K. N. Townsend (eds), *Valuing the Earth: Economics, Ecology, Ethics*, Cambridge, MA and London, England: The MIT Press, 1996 [1971]), pp. 75–88, p. 85.

77   Latour, op. cit., p. 21.
78   A. D. Schrift, *Nietzsche and the Question of Interpretation: Between Hermeneutics and Deconstruction*, New York: Routledge, 1990, p. 71.
79   It should be noted that the ideology should be dissociated from the original thinkers or prophets.
80   Kant brings to bear the objective order of the world upon subjective reason. Yet 'admittedly, the old problem of idealism, how *mundus intelligibilis* and *mundus sensibilis* are to be mediated, returns [in Kant] in a new form' (J. Habermas, 'The Unity of Reason in the Diversity of Its Voices', in J. Schmidt (1996) (ed.), *What is Enlightenment? Eighteenth-Century Answers and Twentieth-Century Questions*, trans. by W. M. Hohengarten, Berkeley, Los Angeles and London: University of California Press, 1996 [1992]; pp. 399–425, p. 401). Subjective idealism as a form of immanent mediation emerges as a conspicuous *instrumentum movendi*.
81   For an in depth-analysis of a good number of binary pairs endemic to Western metaphysics and a proposal of a number of new categories of knowledge by which to subvert the apparent stability of the former see R. Thomas-Pellicer, *What is Culture? The Places of God in an Age of Re-Embodiments*, Newcastle upon Tyne: Cambridge Scholars Publishing (forthcoming); now available as doctoral thesis: 'What is *Kultur?* The Places of God in the Age of Re-Embodiments, University of Surrey, 2012.
82   K. Polanyi, 'Aristotle Discovers the Economy', in K. Polanyi, C. M. Arensberg and H. W. Pearson (eds), *Trade and the Market in the Early Empires*, New York: Free Press, 1971 [1957], pp. 64–94; K. Polanyi, C. M. Arensberg and H. W. Pearson (eds), *Trade and the Market in the Early Empires* (New York: Free Press, 1971 [1957]; G. Baum, *Karl Polanyi on Ethics and Economics*, Montreal and Kingston, London, Buffalo: McGill-Queen's University Press, 1996.
83   T. Jackson, *Material Concerns: Pollution, Profit and Quality of Life*, London and New York: Routledge, 1996, ch. 1; Faber *et al.*, 'Entropy...', op. cit; M. Faber, R. Manstetten and J. Proops with S. Baumgärtner, 'The Use of the Entropy Concept in Ecological Economics' in M. Faber, R. Manstetten and J. Proops, *Ecological Economics: Concepts and Methods*, Cheltenham, UK and Northampton, MA, USA: Edward Elgar, 1998 [1996], pp. 115–35.
84   Jackson, op. cit., p. 12.
85   Highly embodied settlements, or what amounts to the same thing, settlements that run on the mode of re/production.
86   Jackson, op. cit., p. 16.
87   Georgescu-Roegen, *Entropy Law*, op. cit., p. 129.
88   Salleh, 'Embodied Materialism', op. cit., p. 194.
89   Toynbee, *A Study*, op. cit., pp. 50–1.
90   Georgescu-Roegen, *Entropy Law*, op. cit., p. 21; cf. also J. Martinez-Alier, *Ecological Economics: Energy, Environment and Society*, Oxford, UK, Cambridge, USA: Blackwell, 1994 [1990].
91   T. Brennan, *Exhausting Modernity: Grounds for a New Economy*, London and New York: Routledge, 2000, p. 2.
92   T. Berry, 'The Ecozoic Era', New Economy Coalition. Available online at: http://neweconomy.net/publications/lectures/berry/thomas/the-ecozoic-era (accessed 7 September 2014).

Chapter 3

Reclaiming authenticity and ethics

A critique of eurocentric disembodiment and ecofeminist vision of re-embodiment[1]

Madronna Holden

The moral impotence of disembodiment

I once heard a member of the Allied Forces relate his experience liberating a Nazi concentration camp. He found the evil of the camp embodied in the smell of dead and decaying human flesh that penetrated his senses, inciting compassion for the suffering of the camp's inmates. It took weeks after leaving Germany to banish the odour of human agony from his skin and clothes. Yet when he asked villagers living near the camp how they stood the stench, they replied, 'We smelled nothing'. For him, this exemplified the condition under which citizens of the Nazi state accepted genocide on their doorstep. In denying their senses, these villagers also abdicated their conscience. In like fashion, in her post-World War II analysis of totalitarianism, Hannah Arendt observes that totalitarian states consist of individuals 'deprived of seeing or hearing others, of being seen and being heard by them.'[2] Guy Debord adds that such physical detachment allows the justification of oppression by casting others as mere mental representations in a 'pseudo-world' that makes spectators of human beings and spectacles of society.[3]

Indeed, totalitarianism flows all too easily from the Eurocentric worldview with its mind/body split, according to Nobel Prize-winning novelist, philosopher, and member of the French Resistance, Albert Camus. In his novel *The Fall*, the 'absence' from a life lived only 'on the surface' (in a persistent physical fog) allows its protagonist to walk away from the screams of a drowning woman, even as it authorizes self-deceptions regarding colonialism, the obsession with being 'higher' than others, the justification of slavery, and the oppression of women. Altogether, the 'fall' in this novel represents the fall from moral responsibility in European society. Truth cannot be distinguished from a lie, or innocence from culpability, among those detached from lived experience, dividing themselves into parts situated 'both here and elsewhere'.[4] In his non-fiction *The Rebel*, Camus proposes an alternative: the 'new individual' who honours earthly life to the extent that she would neither take the life of another nor her own. Thus

Camus both addresses the question of suicide in the face of the horrors of Nazism and undercuts the justification of violence toward others. For Camus, life can only be authentically lived by embracing it fully, accepting suffering and mortality without otherworldly or ideological redemption. He asserts, 'The earth remains our first and last love; our brothers are breathing under the same sky as we; justice is a living thing.'[5]

The earth is also 'our first and last love' for contemporary ecofeminists who centre their critique of 'dominator societies' (with their multiple oppressions of colonialism, racism, classicism, and sexism) on the interconnected ways humans treat one another and the natural world. Ecofeminists point out that the engine of oppression in such societies runs on the separation of nature from culture and the linking of 'lower' social groups to the 'the sphere of nature' as a biological 'necessity' that undermines the freedom (and too often the very lives) of such groups. The masculinist strains of Judeo-Christian tradition, for instance, objectified forests, streams, and animals – and characterized slaves and women as part of this denigrated 'nature' in order to license the domination of them all. But ecofeminism's cross-cultural analysis indicates that not all human societies are of the 'dominating' type. Indeed, if the natural world, variously related to in various cultures, is the *ground* of human life, it is not its *destiny*. In the variance with which societies relate to the natural world and to other lives, lies the potential for human choice in all societies.[6]

But such choice is stymied by dominating societies' production of the alienated self out of touch both with lived experience and the responsibility for personal choice. In such societies, Charlene Spretnak's 'lone cowboy' and Plumwood's 'warrior-hero… chooses spirit, honor and reputation over life' at the same time that 'he chooses culture over nature,'[7] asserting domination over other humans and the natural world through competition and violence. As Rosemary Ruether observes, this attempt at 'dominating the universe' is also a sign of alienated consciousness, since one can only dominate the world by setting oneself apart from it.[8] Camus' contrasting vision replaces Descartes' 'I think, therefore I am' (which has become emblematic of the mind/body split in the Eurocentric worldview) with 'I rebel, therefore we are'. Authenticity exists in the rebel's making her own decisions as against law, social role, ideology, or mental abstractions imposed from without. The 'new individual' is a central participant in the 'we are', the community built on the premise that 'I have need of others who have need of me and of each other.'[9] This individual parallels ecofeminist Plumwood's 'ecological' or 'mutual' self who

> salutes the social order as another self, a center of subjectivity like mine but a different one, one which imposes a limit on mine, and incorporates this into the concept of the I. Similarly, the ecological self recognizes the agency or intentionality having its origin and place like

mine in a community of the earth, but as a different one... which limits mine.[10]

Thus the authentic self is also the moral self bound by the reciprocity and limits created by the presence of others: indeed, without such 'discipline', to use Camus' words, 'the individual... is only a stranger, bowed down under the weight of an inimical collectivity.' This 'inimical collectivity' is expressed in the antagonism between individual and society in any society that 'loses its direction' by denying the interdependence or 'we are' of its members.

If Camus' authentic individual is 'new' to the contemporary Eurocentric state, it is old to enduring indigenous societies. Following the traditions of indigenous societies who see themselves as members of an earthly family of natural life, Ruether proposes we recast the *I-it* relationships of capitalism, industrialism, and colonialism into *I-thou* relationships.[11] Self-termed 'geologian' Thomas Berry concurs, countering the destructive tendencies he sees in patriarchal Eurocentric societies by asserting that relationships to all 'others', human and more than human, should be 'a communion of subjects rather than a collection of objects.' Berry discusses Native American history both in terms of the necessity of correcting the legacy of colonialism in US society and the models certain native traditions present of earth-centred and non-dominating societies.[12] Indeed, Camus' interdependent 'we' and Plumwood's 'mutual self' exhibit the self-same interweaving of individual autonomy and individual responsibility as Dorothy Lee sketches for the Navajo and Lakota, and I have detailed for the Chehalis.[13] Jacob Bighorn, former administrator of the Chemawa Indian School in Oregon put it this way: it was the traditional responsibility of the adults of Native Northwestern communities in what is now the US to support a child in finding her distinct gift, which she could then give back to life. Notably, the path in finding an individual gift could only be known – just as it could only be lived – by the child herself.

In such social contexts, moral responsibilities flowing from one's embodiment and embeddedness in the natural world seem apparent. From 'sharing one skin' – the Okanagan (in the US Pacific Northwest) term for community – flow compassion and care.[14] As Nisqually Tulalip elder Janet McCloud (from US Puget Sound in Washington State) puts it, our body *is* both nature and 'our first teacher' – modelling the proper relationship to other natural life, human and more than human. Without such a relationship to the natural world, we live lives of 'spiritual poverty' comparable to that of Camus' 'stranger' who suffers both physical displacement and moral detachment.[15]

A story recently circulated in native oral tradition in the US Northwest entails a query put to a native elder as to how non-Indians could help Native people. The non-Indian received a pointed reply: 'You could stay in one place.' These words are a distinct jab at the dynamics of the colonialism that displaces indigenous lives on the lands it overruns, even as it displaces the agency of those it relegates to the status of 'objects' (nature, body, women, and indigenous peoples) with those it licenses as subjects (humans, mind, men, and 'civilization').[16] Indigenous rootedness stands in profound contrast to such displacements. When I first visited Lower Chehalis elder Henry Cultee in US Washington State, he was eighty-four years old and living in his 'fishing shack' on Grays Harbor, at the place called *Samamanauwish* – which was also Cultee's inherited name, following the tradition by which his people named themselves for their places on the land. His personal power, to which he attributed his own long life, means 'rooted to this ground'. As a child, he dived in the local river to share its own long life. In training for this practice, his elders told him: 'Don't lie to your life: the eyes of the world are looking at you'. This statement succinctly links personal authenticity (being truthful to one's life) with moral behaviour (entailing generosity, reciprocity, and care) exacted by the 'eyes of the world' that watch over human actions, and according to Chehalis tradition, apportion the length of human lives accordingly.[17]

Pioneer displacement held them to no such ethical standards. Without a sense of earthly belonging – including that most intimate of belongings, to one's own body – humans were 'without a heart', according to Okanagan teacher Jeanette Armstrong. This blinded them to the destructive effects of their actions on the world from which they detached themselves.[18] Expressing a similar perspective, Chehalis mothers traditionally saved their umbilical cords for their children – since a child lacking an intimate connection to body/earth/mother would run around in a hyperactive way, unable to listen to others.[19] Such 'displacement anxiety', as Armstrong aptly terms it, goes hand in hand with a hunger that 'eats up' other lives for its own purposes – relegating other species and whole native nations to the land of the dead in a story Henry Cultee told me.[20]

Certain pioneer family descendants I interviewed in Western Washington in the 1970s expressed their own ambivalence over their ancestors' pervasive displacement. 'We are a restless race', one woman told me, relating her ancestor's habit of abandoning his children to take off on one 'scheme' after another. In his letter to Cornwall, England, from Grand Mound, Washington in 1860, pioneer Samuel James observed: 'The Americans… generally calculate to… do a little work on a piece of land, and then watch the first opportunity for selling… and thus the great multitude of them are always on the move.'[21] Those who 'were always on the move' treated the land and its inhabitants as a 'one-night stand', to use contemporary ecologist Wendell Berry's words.[22] They did not stay around to witness the results of their actions but simply took what they wanted and went on their way. Henry Cultee averred that whites engineered some useful things, but they had a tendency to 'chew right through a mountain rather than go around.' Native memory in Western Washington is scarred by the ways in which pioneers consistently burned native houses to gain the land on which they stood, even as they declared war on other species who stood in their way – wolves, mountain lions, grizzly bear, coyote, and even elk and geese who consumed emigrant crops.

Backgrounding and perceptual denial

US pioneers sidestepped the considerable moral concern of such actions with the narrative of Manifest Destiny, in which native peoples faded away before the march of pioneer progress. Native peoples certainly faded away before pioneer *vision*. Such 'invisiblizing', which Val Plumwood also terms 'backgrounding' or 'systematic non-seeing',[23] transformed the land and its lives from subjects with agency and intentionality into objects that served pioneer goals. Pioneers experienced a concomitant disembodying process as their plans for the future of the land overrode the evidence of their own eyes. Thus an early missionary saw the fertile lands tended by the Chehalis (including their villages with 200-foot-long potlatch houses used for intertribal winter ceremony) as 'wastes' inhabited by the 'lonely huts of the Indians.'[24] In like fashion, an early explorer saw the Kalapuya of the US Willamette Valley as a people who simply lived out under the trees – overlooking entirely the sixty-foot-long plank houses that constituted their winter village and held the stored food with which these same Kalapuya began saving pioneers from starvation three decades later.[25]

This dynamic, in which the mind overruled the eyes, is echoed in the 'fantastic' observational errors concerning women's bodies that James Hillman delineates in the history of Western philosophers, psychologists, and scientists who set themselves up as 'observers' of the female body they saw as 'datum'. As Hillman remarks, if seeing is believing, believing is also seeing. These observers thus 'saw' the 'structural inferiority of the woman's body' in which they believed. Hillman concludes that Western consciousness will never be whole nor recover its connection to the *animus mundi*, the soul of the living world, until it heals this history of misogynous and mistaken perceptions of female bodies.[26]

As did the scientists who saw in women's bodies what was never there and pioneers who failed to see in native villages what was there, Nazi doctors muted their senses and their consciences together in the 'psychic numbing' that enabled them to carry out their atrocities. Robert Jay Lifton, who interviewed such doctors, found that in one part of their lives – the one in which they were situated in their bodies – these doctors upheld their responsibilities as fathers and husbands as well as doctors. But in their work in the concentration camps, they set mind over matter, 'psychically numbing' themselves. They well understood that had they been present to the touch, smell, sight, and sound of the pain they caused, they would not have been able to continue their actions. Lifton observes that the Nazi doctors were not alone in expressing the egregious potentials of disembodiment. He gives examples of US scientists, psychologists, and doctors who also practise 'psychic numbing' to suspend their ethics in the treatment of patients and carry out questionable research experiments.[27] We might add to this list the Chief Economic Officers of major chemical companies in the 1950s, who viewed the x-rays showing their workers' bones dissolving from exposure to the chemicals they worked with and formed an international consortium to keep this information hidden in order to protect their profits.[28] Lifton's proposal for immunizing society against the moral dangers of such psychic numbing is twofold. First, he suggests that professionals prioritize an embodiment of their knowledge and experience. Second, he suggests that empathy for others be a guiding rule in professional practice.

Mental separations versus embodied sharing

For their part, US indigenous Northwesterners such as the Coast Salish rejected pioneer objectifications of themselves by constructing satirical portraits of their objectifiers.[29] In response to a scholar's seeking linguistic examples of 'objects' in the Pit River language, elder Wild Bill of US California asserted that he could only understand 'objects' as 'dead people'. In turn, those who 'don't believe anything is alive', who dismiss the vitality of the material world, 'are dead themselves.'[30] This echoes Lifton's insight that objectifying others is commensurate with deadening oneself. Andrea Nye further exposes the double objectification of self and other in her feminist analysis of the history of Western logic. Beginning with Parmenides and his rejection of bodies in general and women's bodies in particular, and ending with twentieth-century Nazi sympathizer Gottlob Frege, Nye discusses the thinking that makes of 'natural life... only dead skin that falls away before the hard bone of logic... unrelated to conflictual relations between men and women, between men, or between men and the natural world'. Such thinking allows one to be 'a dogmatist in religion, a fascist in politics, a sadist in love' – all supposedly irrelevant to one's logic. With the collapse of his own subjectivity (and participation in intersubjectivity) and his reduction of human communication to a mathematical formula, Frege reaches the ultimate endpoint of disembodied logic 'out of touch with all reality... the reality in the streets, the disappearances, the deaths, the concentration camps, out of all touch with others, who... think like him because they must, like he must, think... [like] a thing.' In its arrival at the 'thinghood' of self and other, Frege's 'logic, in its final perfection' abandons sanity along with ethics.[31]

'The insanity of such systems', Arendt writes in turn, 'lies... in the very logicality with which they are constructed... the curious logicality of all isms' that 'harbors the first germs of totalitarian contempt for reality'. These intellectual systems resolutely empty the world of everyday 'common sense' in order to impose on it a 'supersense', an abstract 'solution to the riddles of the universe'. Such thinking yields the logic useful to totalitarian regimes – in which 'everything follows' not only 'comprehensively' but 'compulsively'.[32] This is what Plumwood terms the 'logic of colonialism' and what the Okanagan term as 'talking talking inside the head' severed from body, Earth, and community.[33] To affect such logic one must not see what one's eyes see. And certainly, one must not feel what others feel in 'sharing one skin'. Such logic can only be maintained in the rarefied air of the disembodied mind that denies the body's evidence.

This is fully in line with the capitalist notion that one must 'earn a living' – that the body in itself is undeserving of food and shelter. In contrasting indigenous contexts, to be born in a human body is to be born with rights to shared sustenance, as illustrated by the prevalence of food sharing throughout the US Pacific Northwest, which sustained pioneers when they first arrived.[34] Henry Cultee stipulated that one shouldn't even ask, 'Are you hungry?' when a guest arrived. One should just bring out the food – lest there be any intimation that food was under the control of an individual choosing to dole it out.

Consciousness of shared embodiment prompts us to make sure that our neighbours are fed if we are. By contrast, as Sylvia Frederici and Carolyn Merchant document, Eurocentric disembodiments have historically destroyed the occasion for such sharing, even as they have justified imposition of physical punishment such as forced labour and obedience to coerce behaviour.[35] Thus the widespread violence toward women in the contemporary world, the massacres perpetrated on indigenous peoples in global colonial history, and the physical punishment of children to socialize them derive from a common root: a worldview that authorizes the mortification of the body as a means of social control. The concept of original sin, first proposed by Augustine during the period in which institutionalized Christianity became entwined with empire, is a case in point. This doctrine specifying that all are born with the stain of sin on their souls made women's bodies especially culpable, since it was through them that original sin was passed at birth. Original sin's assertion of the flawed character of earthly life rationalized both human suffering and authoritarian social arrangements – and urged the acceptance of both.[36] This conceptualization divided the world into 'the suffering body deprived of agency, and the mastering external rational agent.'[37]

Enacting this worldview, US missionaries and boarding school administrators used beatings, hunger, and forced labour on native children with the rationale that punishing their bodies was the cost of saving their souls.[38] In an example from the present day, psychic numbing and physical numbing of men's bodies are coincident with one another in team sports –

which teach boys to turn their bodies into instruments with which to injure other bodies. Michael Messner exposes the irony with which those who originally become athletes to honour their bodies learn to objectify them instead as they 'play through the pain.' Many take up drug and alcohol use to numb themselves to their body's signals – and push them beyond their natural limits. Not incidentally, Lifton found that the Nazi doctors also used heavy drinking to maintain their numbness.[39]

Dissecting nature and women's bodies

Mortifying the body as a means of social control goes hand in hand with science's punishing technologies for controlling the natural world. Ecofeminists have observed the pointed similarities between the experimental methodology of emerging science during the Renaissance and the torture of the witches, whose murders by the millions went hand in hand with the rise of capitalism.[40] Francis Bacon, father of the modern scientific method, used the language of the witch trials to justify the forced interrogation of nature – not surprisingly, given his profession as trial lawyer during these same trials. Bacon's assertion that any procedure that leads to nature's revealing her 'secrets' is good *per se* has an insidious side, as well as insidious persistence. Hillman notes that nineteenth-century French psychiatrists used a diagnostic method on women patients formerly used by the Inquisition on witches.[41] In his *Gendered Atom*, Theodore Roszak exposes the Eurocentric worldview that turns its objectifying gaze similarly on women and nature – and licenses the rape of them both.[42]

Scientific dissection of a once-living body to force it to reveal itself rests on the illusion that a dissected body is the equivalent of one that is alive, whole, embedded in a living system, and imbued with agency. The notion that dissection leads to knowledge of the dissected life is false in the same way as is Freud's assertion that he understood the consciousness of the girls whose bodies he objectified and thus faultily observed.[43] The scientist in such a scenario views natural life as emptied of its story in the same way that a plastic-wrapped cut of meat in a supermarket case is emptied of the story of its animal life. An examination of such packaged meat yields no knowledge of the singular being whose flesh it is, indicating only the distance between the consumers' own animal body and the animal body which sustains him or her.

Science's dissection of the bodies of natural lives to expose their 'secrets' parallels social exhibition of the bodies of denigrated social groups. In Lisel Mueller's poem, 'The Unanswered Question', she asks us to give up our affiliation with the curiosity seekers who came to stare at the last survivor of her Tasmanian culture put on display in a cage in London and stand instead in bodily sympathy with the displayed woman, pondering our responsibility for understanding any word of her lost language that might pass the bars between us.[44] The display of the Tasmanian woman is doubly egregious as an example of what Maria Mies terms the 'white man's dilemma', in which members of colonializing cultures seek to recover their lost bodily feeling through the experience of exotic places and peoples – even as they continue to destroy them.[45] Camus reverses this process in *The Fall* by putting Eurocentric male consciousness on exhibit, unmasking both the pettiness of its impulse to dominate and the self-deceptions in the search for sensual recovery in artefacts garnered from colonial outposts.[46]

The oppression entailed in the exhibition of other bodies is illustrated in John Stoltenberg's workshops, in which men assume the poses of women exhibited in pornographic magazines. Even though they remained fully clothed, as these men gained bodily empathy with exhibited women through sharing their postures, they expressed feelings of violation and humiliation, and the 'alienation affect' flowing from being put on exhibit, which Stoltenberg describes as the feeling of having 'my head... detached from my body' and face 'put on'. Stoltenberg also notes that in the ten years of conducting these workshops with hundreds of men, those who excused the demeaning nature of women's exhibition since, 'they got paid for it', were never men of colour. To a man, men of colour in these workshops sensed the exploitation involved in the women's exhibition.[47] In turn, the exhibition of women in contemporary pornographic magazines follows a centuries-old tradition in Western art that depicts women as arranged before the appropriating gaze of the viewer, which emptied the exhibited women of their agency as surely as the colonial gaze emptied indigenous lives and landscapes of theirs.[48]

Echoing scientific dissections, contemporary media often atomizes women's bodies, portraying them as consumable in the same way that capitalism portrays nature as consumable. In *Deadly Persuasion*, Jean Kilbourne observes contemporary media's depiction of women's bodies as things or parts of things – legs, buttocks, headless torsos – or feet as in the notorious collection of ads showing part of a women's body protruding from a garbage can, wearing an advertised pair of shoes. Media objectification of women also portrays silenced or voiceless women, girls as young as four or five years of age as sex objects, and women's bodies as passive objects of violence. Kilbourne quotes one young woman's response to such ads: 'I saw myself piece by piece. I didn't see the connections... I didn't see *me*... I hated my body because I hated myself. I doubted my body because I doubted myself.'[49] In the wake of her personal struggle with an eating disorder, Naomi Wolf compares media images of women's bodies to the 'iron maiden' used to reshape women's bodies as a torture device in the Middle Ages.[50]

Kilbourne labels such media depiction of women's bodies as 'killing', echoing Pit River elder Wild Bill's point that objectification results in 'dead people' and Carolyn Merchant's historical analysis of the 'death of nature' with the objectification of the natural world in industrialization – a profound contrast with the 'participatory' consciousness of earlier embodiment.[51] In the contemporary context, Kilbourne notes that media images of women's bodies are growing more phantom-like in their passive immobility, their airbrushed stylization and absence of pores. Stuart Ewen's comparative chart of the media images of the female body over the past few decades shows how women's bodies have been literally disappearing – which we might surmise from the new size 'zero' clothing. 'When your body says more', one commercial text cited by Ewen states, 'Say less'. Images of female voicelessness, including ads in which women's mouths are covered up or even sewn shut go hand in hand with socialization resulting in girls' failure to 'speak up for themselves or to use their voices to protect themselves'.[52] Here women of depth and flesh are missing as the proffered image cannot be emulated short of Parmenides escape to the realm of pure concept or Frege's logical insanity. In fuelling consumption, however, such images are successful in their failure. Each time one product fails to deliver, as it inevitably does, another arises to take its place. As real women lose agency in this process, the engine of consumption gains momentum. The fact that many women are literally starving themselves to death while the US over-consumes to the extent that its five per cent of the world's population consumes a quarter of its resources exposes the connections between the oppression of nature and of women. As explored in Katherine Gilday's *The Famine Within*, women's bodies have become the symbolic battleground for the control of nature in an arena in which women's bodies and nature, both of which nourish others, have no rights to nourishment themselves.[53]

Visions of re-embodiment: physical vulnerability and transcendence through the body

The flight from the body in Eurocentric philosophy has historically been a flight from physical vulnerability, as Plumwood elaborates in her discussion of Plato. But as she observes, the failure to accept death is also a failure to accept life in setting other-worldliness or ideological transcendence over lived existence. Notably, such transcendence has historically been linked with militarism and the ideal of the warrior who 'in risking death demonstrates his control over and disregard for the body'.[54] Indeed, to idealize or romanticize the body, ignoring its vulnerability and imperfections, is to enter the realm of dualism that substitutes conceptual perfection for reality; surfaces and images for depth and complexity. As Nicaraguan poet Daisy Zamora tells us in her 'Celebration of the Body', to celebrate the body in which we each live a life is also to accept that that body 'can hurt and get ill.'[55] It is our fragile skin, capable of being wounded, which allows us to touch and be touched.

Further, our physical vulnerability has played a central role in the creation of human culture. Even as the physical vulnerability of the young provides the occasion for sharing cultural values and historical lessons, the vulnerability of the elderly provides the occasion for receiving their experience and knowledge. Moreover, there is distinctive resilience to the community that understands what its most vulnerable members indicate about the need for care and repair. As the authentic presence in a body teaches compassion for the suffering of women subject to domestic violence, for instance, that violence becomes unacceptable and a commitment emerges to change the social conditions that foster it. Physical compassion for those stricken by cancer can mobilize a society to prevent the chemical pollution that is centrally implicated in the cancer epidemic. Likewise, responding to the physical vulnerability of the youngest among us impels us to clean up the toxins now endemic in human embryonic fluid and mother's milk. Understanding our physical vulnerability is also an underpinning of the precautionary principle, which indicates that one should not release anything into an ecological system until it is proved safe to other lives.

Those who share the suffering of others and help repair the conditions that create it also foster their own well-being in an interdependent world, in which vulnerable bodies rely upon one other for meeting physical and emotional needs. Thus we may understand Camus' words, 'I have need of others who have need of me and each other' as characterizing a community of those who accept the vulnerabilities that make each of us both unique and open to others.

Authentic embodiment teaches humans their place in the natural order, in the reciprocal balance that interweaves all lives – and our own turn at life with others. As 'Godfather Death', a folktale told throughout the world relates, human embodiment entails facing the limits of human lives.[56] In Eurocentric thinking, transcendence has revolved around escaping such natural limits, from Parmenides' flight from mortality to Christianity's other-worldliness, to science's attempts to control and reshape nature by setting humans above it. But as we have seen, attempts to escape our place in natural cycles by whatever means are fraught with hazards in the treatment of other humans, as well as the treatment of more than human lives. Thus I would like to propose transcendence *through* rather than *escape from* the body. This idea of transcendence is based on a cultural model in which the hallmark of human maturity is the extension of self to others, in a dialectical alternative to the closing off of the territorial self in Eurocentric cultures.[57]

In such transcendence, we extend our lives as far into the past as we remember, and as far into the future as we are remembered by the imprint of our actions. Such transcendence sets each of us in dialogue with other times and other lives as an alternative to the impulse that would escape time and space. In a transcendent practice common among indigenous peoples, human individuals form an alliance with a being in the more-than-human world in order to expand their humanity, as exemplified by the title of Deborah Rose's work on the Aboriginal Australians, *Dingo Makes Us Human*.[58] Many traditional Chehalis stories, as do folktales throughout the world, take place in the time 'when all the animals were people': their tellers embody each of the story's characters, facilitating the audience's entry into the arena in which these 'others' have families, homes, and feelings like their own. Even as Jeanette Armstrong speaks of the body that connects us with one another in compassion, Val Plumwood cites the words of an elder in Australia: 'my body... is the "same your body."'[59] Thus we feel that when the land is ravaged, in Armstrong's words, "It is my body that is being torn, deforested, and poisoned by development"'.[60]

Visions of re-embodiment: affiliation and desire

Authentic embodiment consists not only in paying attention *with* the whole body, but *to* it. Attending to the body's messages concerning hunger has the power to heal the unhealthy relationships with consumption encouraged by contemporary advertising. Authentic re-embodiment also exposes the illusion that consuming the pain other beings raised on factory farms is adequate nourishment for bodies we honour. It tells us that the lives that sustain ours, embodied as we ourselves are, deserve reverence, care, and thanksgiving, as well as a full life lived according to the requirements of their species.

Thus authentic re-embodiment requires a shift away from the technology of control, segmentation, and mechanization. Hillman observes that in an anaesthetizing environment of 'plastic, Styrofoam, cold metal', his male patients easily became 'brutal'.[61] Such anaesthetization inhibits bodily relationships with others just as technology today too often separates consumers from the sensory and ethical cognition of the products they consume. In plucking a chocolate bar from a grocery store shelf, its consumer has no experience of the child enslaved on an African planation (as many are today) to harvest that chocolate. Between biting into the chocolate and the growing of its ingredients lie such complex lines of transportation, reshaping, and packaging that there is neither the recognizable earth of Africa nor hands of the harvester left for the one eating it to perceive.

By contrast, the Chinook at the mouth of the US Columbia River saw their world as vitally 'alive' – and created their technology accordingly.[62] They carved a heart and lungs on their ocean-going canoes, sailing them as a living body working with a living body. Here is a challenge authentic re-embodiment must meet: to replace Bacon's technology of control (which bears too many affiliations with the technology of torture) with a technology that expands its users' sensual reach to other bodies.

Further, any vision of authenticity must address and counter the constraining roles that make being born into a particular body a liability under hierarchical social arrangements. In the case of women, for instance, 'the female body itself comes to seem oppressive' when 'women's agency is denied'. Yet we have seen the ethical hazards of abandoning our own bodies as would those who assert that liberation entails escaping female bodies through technological alternatives to child-bearing or joining men in the arena of pure rationality.[63] Gender ground provides an alternative even as it counters the three related Eurocentric displacements: displacement from body, from community, and from land. It allows us to inhabit our place on Earth and in society as we do our bodies. Gender ground is a place where women and men and those who choose to walk between these (as 'bridge' or flexibly gendered persons) stand in affiliation with one another. Gender grounds are embracing rather than exclusive or preordained by tradition or convention. Such gender grounds have a physical location in some cultures. And in most indigenous cultures, they also have a mythic dimension, imparting meaning, authority, and responsibility to being a woman, man, or flexibly gendered person. For instance, an Apache girl connects with Changing Woman in her puberty ceremony, thus also connecting with the power of creation as she enters the gender ground of her womanhood.

Gender ground is also generational ground onto which elders welcome the young to guide, care for, protect, and encourage them. On such ground, Malidoma Somé relates how a Dagara 'male mother' (a mother's brother in his matrilineal society) in West Africa, worked to 'take the anger out of a young man' (in this instance, himself) caught between the white and the traditional Dagara world.[64] Peggy Orenstein observed the contrasting situation of girls in the contemporary US in the absence of a circle of women elders to welcome them onto a ground where they might find protection, nurturance, and purpose – and honour for growing into the body of a woman. Instead, Orenstein found that adolescent boys initiated these girls into womanhood, grooming them for compliance, and shaming them for personal success, such as academic achievement. These girls' mothers were both isolated from one another and constrained in their roles within the nuclear family – and comparatively helpless to offer their daughters meaningful support. The girls thus became young women alienated from their bodies and lacking personal purpose, in contrast with the Apache girl whose initiatory dance is a powerful prayer for the whole of her people.[65]

Altogether, re-embodiment allows the reclamation of authentic desire, of desire as personal vision as against the manipulated desire that consumerism hawks as bizarrely disembodied sexual images or products to fill unmet needs for meaning or connection. Our authentic presence in our bodies tells us that our desire for such things as clean air, fresh water, untrammelled and toxin-free natural landscapes, and a vital world to pass on to succeeding generations, cannot be bought or sold. Authentically embodied desire is a steadying rudder to steer us to those acts that no one else can do in our stead. For each of us in a body, such desire is an intimate expression of who we are – and who we might be.

Warm Springs elder Lizzie Pitt from Eastern Oregon eloquently expressed her own full-bodied desire in words depicting the interdependence of all natural life: 'Someday the land will be our eyes and skin again.'[66] When such a day comes, the re-integration of body and mind, self and other, human and nature will come with it. Then all of us will be able to celebrate together the shared gift of our embodied lives.

Notes

1   Another version of this essay appears as 'Coming Home to Our Bodies/ Healing the Earth We Share', in P. Godfrey and D. Torres (eds), *Emergent Possibilities for Global Sustainability: Intersections of Race, Class, and Gender,* vol. 2, London and New York: Routledge, forthcoming.

2   H. Arendt. *The Origins of Totalitarianism,* New York: Harcourt Brace Javonovich 1973, pp. 457–8.

3   G. Debord. *Society of the Spectacle* 1977 trans. Red and Black. Available online at: www.marxists.org/reference/archive/debord/society.htm (accessed 6 November 2014), pp. 191–203.

4   A. Camus, *The Fall,* trans. Justin O'Brien, New York: Vintage 1956; pp. 23–25, 44, 49–50, 102–105.

5   A. Camus, *The Rebel,* trans. Anthony Bower, New York: Vintage 1956; pp. 11, 296–7, 306.

6   M. Mellor, *Feminism and Ecology,* New York: New York University Press 1997, pp. 7–13; M. Mies and V. Shiva, *Ecofeminism,* London: Zed Books 1993, pp. 98–163; V. Plumwood, *Feminism and the Mastery of Nature,* London and New York: Routledge 1993, pp. 3–39; 44–7, 79–81, 109–16; V. Plumwood, 'Nature in the Active Voice', *Australian Humanities Review,* 2009, 46, pp. 118–20; A. Salleh, *Ecofeminism as Politics,* London and New York: Zed Books Ltd, 1997, p. 36.

7   V. Plumwood, 'Inequality, Ecojustice and Ecological Rationality', *Ecotheology,* 1998–99, vols. 5 and 6, p. 96; C. Spretnak, *The Resurgence of the Real, Body, Nature and Place in a Hypermodern World.* Reading, MA: Addison-Wesley, 1997, p. 71.

8   R. Ruether, *New Woman, New Earth,* New York: Seabury Press 1975, p. 148.

9   Camus, *Rebel,* p. 297.

10   Plumwood, 1993, p. 129.

11   Ruether, p. 148.

12   T. Berry, *Evening Thoughts, Reflecting on Earth as Sacred Community,* San Francisco, CA: Sierra Club, 2006, p. 96; T. Berry, *The Dream of the Earth,* San Francisco, CA: Sierra Club 1988, p.138–62.

13   D. Lee, *Freedom and Culture,* Englewood Cliffs, NJ: Prentice-Hall, 1959, pp. 5–14, 59–69; M. Holden, 'The Realization of Self in Primitive Society: The Example of the Chehalis Indians', *Dialectical Anthropology,* 1996, vol. 20, pp. 1–43.

14   J. Armstrong, 'Community: Sharing One Skin', in J. Mander and V. Tauli-Corpuz (eds), *Paradigm Wars: Indigenous Peoples' Resistance to Economic Globalization*, San Francisco, CA: Sierra Club Books, 2005, pp. 35–9.

15   J. McCloud, 'On the Trail', in J. White (ed.), *Talking on the Water*, San Francisco, CA: Sierra Club Books 1994, p. 252.

16   Plumwood, 1993, pp. 25–9, 32–3, 43–4, 51–61, 80; 1998–9: pp. 185–218; 1999: p. 167; M. Mies. *Patriarchy and Accumulation on a World Scale*, London: Zed Books, 1986, pp. 64–5; Salleh: p. 36; Ruether: p.198.

17   M. Holden, 'Re-storying the World: Reviving the Language of Life', *Australian Humanities Review* 2009, vol. 47, pp. 141–57.

18   Armstrong, 2005. p. 38; J. Armstrong, 'Keepers of the Earth', in T. Roszak, M. Gomes and A. Kanner (eds), *Ecopsychology*, San Francisco, CA: Sierra Club Books, 1995, pp. 317–8.

19   T. Adamson. 'Sources of Chehalis Ethnology', MS in Melville Jacobs Collection, University of Washington Library 1926, pp. 224, 254.

20   M. Holden, 'Making all the Crooked Ways Straight: the Satirical Portrait of Whites in Coast Salish Folklore', *Journal of American Folklore*, 1976, vol. 89, p. 263.

21   D. James. *From Grand Mound to Scatter Creek*, Olympia: State Capitol Historical Association of Washington 1980, p. 39.

22   W. Berry, 'Back to the Land', in *The Amicus Journal*, 1999, vol. 20:4, pp. 37–40.

23   Plumwood, 1993, pp. 4, 21–2, 47–8, 54–70, 123.

24   C. Landerholm, (ed. and trans.), *Notices and Voyages of the Famous Quebec Mission to the Pacific Northwest*, Portland, OR: Champoeg Press for the Oregon Historical Society, 1956, p. 39.

25   H. Mackey, *The Kalapuyans*, Salem, OR: Mission Mill Museum Association and Harold Mackey 1974, pp. 2–3.

26   J. Hillman, *The Myth of Analysis*, Chicago, IL: Northwestern University Press, 1972, pp.238–43.

27   R. J. Lifton, *The Nazi Doctors: Medical Killing and the Psychology of Genocide*, New York: Basic Books, 1986, pp. 141, 419–21, 442–5; 501–4.

28   B. Moyers, *Trade Secrets: A Moyers Report*. Available online at: wwwpbs.org/tradesecrets/ (accessed 18 March 2013).

29   Holden 1976, pp. 271–93.

30   B. Callahan (ed.), *A Jaime de Angulo Reader*, Berkeley, CA: Turtle Island Press, 1979, pp. 240–1.

31   A. Nye, *Words of Power: A Feminist History of Logic*, New York and London: Routledge, 1990, pp. 1–5; 163–71.

32   Arendt, p. 458.

33   Plumwood, 1993, p. 41; Armstrong 1995, pp. 319–20.

34   E.g. F. Boas, 'Chehalis Folklore', manuscript, pp. 865–70, 881–3; Adamson, pp. 4, 84, 345; L. Youst and W. R. Seaburg, *Coquelle Thompson, Athabaskan Witness, A Cultural Biography*, Norman, OK: University of Oklahoma Press, 2002, pp. 62–4; R. Thomas, 'Life in the Siuslaw Valley prior to European Settlement: Glimpses of Tribal Life Styles at the Convergence of the Siuslaw, Kalapuya, Yoncalla, and Lower Umpqua Tribal Domains', Eugene, OR: University of Oregon, M.A. thesis: 1999, pp. 6–15; J. Thornton, compiler, *The Indians of the Oregon Coast, The Ancient and Original Inhabitants, as Recorded in the*

John P. Harrington Collection of the Smithsonian Institution National Anthropological Archives, cited in Coos Bay, OR: Indian Education, manuscript, no date, p. 19.

35   S. Federici, *Caliban and the Witch*, New York: Autonomedia, 2004; S. Federici, 'Capital and the Body; A Rejoinder to Salvatore Engel-Di Mauro' *Capitalism Nature Socialism*, 2006, vol. 17:4, pp. 74–7. C. Merchant, *The Death of Nature*, San Francisco, CA: Harper and Row, 1980, pp. 132–43.

36   M. Fox. *Original Blessing*, Santa Fe, NM: Bear & Company, 1983, pp. 48–51.

37   Plumwood, 1993, p. 38.

38   D. W. Adams. *Education for Extinction, American Indians and the Boarding School Experience, 1875–1928*, Lawrence, KS: University Press of Kansas, 1995; S. Reddick, 'The Evolution of Chemawa Indian School: From Red River to Salem, 1825–1885', *Oregon Historical Quarterly*, 2000 , num. 101, 442–64; E. Chalcraft, 'Memories Storehouse' 1888, (trans. Helen Neilson, held in Washington State Library manuscript collection).

39   M. Messner. *Power at Play*, Boston, MA: Beacon Press, 1992, pp. 64–76, 93–102; Lifton, pp. 193–4; 443–4.

40   Federici, 2004; Federici, 2006, pp. 74–7; M. Mies and V. Shiva, p. 134; Merchant, pp. 132–43.

41   J. Wishloff, 'Spe Salvi: Assessing the Aerodynamic Soundness of the Civilized Flying Machine', *Journal of Religion and Business Ethics*, 2009, num. 1, pp. 3–7; S. Frederici, 2004, pp. 200–6; Spretnak, pp. 54–5; Hillman 1972, p. 255.

42   T. Roszak, *The Gendered Atom: Reflections on the Sexual Psychology of Science*, Berkeley: Conari Press, 1999.

43   Hillman, 1972, pp. 238–43.

44   L. Mueller, 'The Unanswered Question', in *Alive Together*, Baton Rouge and London: Louisiana State University Press, 1996, p. 11.

45   Mies and Shiva, pp. 132–63.

46   Camus, *The Fall*, pp. 44–48, 67–69, 72.

47   J. Stoltenberg, *What Makes Pornography 'Sexy'?*, Minneapolis, MN: Milkweed Editions, 1994, pp. 14–7, 28–30, 54–56, 72–4.

48   J. Berger, *Ways of Seeing*, New York and London: Penguin, 1972.

49   J. Kilbourne, *Deadly Persuasion*, The Free Press: New York, 1999, p. 259.

50   N. Wolf, *The Beauty Myth*, New York: Doubleday, 1991, pp. 17–8, 51, 73, 135, 228, 266–9, 278–80.

51   C. Merchant, *Ecological Revolutions*, Chapel Hill and London: University of North Carolina Press 1989, pp. 19–26; 198–251; 258–60.

52   Kilbourne, op. cit. and *Still Killing Us Softly 3: Advertising's Image of Women*. Available online at: www.youtube.com/watch?v=_FpyGwP3yzE (accessed 18 March 2013); S. Ewen, All Consuming Images, New York: Basic Books, 1988, pp. 180–3.

53   K. Gilday, writer and director, The Famine Within (film). Kandor productions, 1990. P. Orenstein, *Schoolgirls*, New York: Doubleday, 1994, pp. 94–102.

54   Plumwood, 1993, pp. 84–103.

55   D. Zamora, 'Celebration of the Body', quoted in B. Moyers, *The Language of Life*, New York: Doubleday, 1995, p. 440, from *Life for Each*, bilingual edition, trans. Dinah Livingstone, London: Katabasis, 1994.

56   M. Holden, 'Godfather Death', *Parabola*, 2002, vol. 27:2, 22–6.

57   Holden 1996, pp. 1–43.

58 D. Rose, *Dingo Makes Us Human*, New York, Cambridge University Press, 2000.

59 Plumwood, 1993, p. 139.

60 Armstrong 2005, pp.35-9.

61 J. Hillman and M. Ventura, *We've Had a Hundred Years of Psychotherapy and the World's Getting Worse*, San Francisco, CA: HarperSanFrancisco 1992, p. 212.

62 STOWW (Small Tribes of Western Washington) treaty rights workshop handout, Olympia, Washington: 1975.

63 Mellor, pp. 7–8; Plumwood, 1989, p. 36.

64 M. Somé, *The Healing Wisdom of Africa*, New York: Jeremy Tarcher, 1997, pp. 8–9.

65 Orenstein, pp. 54–8, 62–5, 68–87, 94–6, 237–8, 269.

66 C. Stowell, *Faces of a Reservation*, Portland, OR: Historical Society Press, 1987, p. 104.

Part 2

The sacred dimension

Chapter 4

Towards a deconstruction of leadership and cosmology

The re-embodiment of the sacred

Ali Young

Apocalypse or bust?

Over the last twenty years or so, postmodern theoretical contributions have done a great deal to highlight the importance of subjectivity in the construction of knowledge. Indeed they have been part of a crucial process that has, ultimately, resulted in the erosion of the unquestionable legitimacy of hegemonic discourses that seek to present themselves as 'truth'.[1] Within postmodern frameworks, knowledge is 'grounded in human society, situated, partial, local, temporal and historically specific'.[2] It is produced through the emotional, sexual, embodied, gendered experience of human beings. As such, knowledge is inextricably entwined with political process, contestation, conflict and change.

In this chapter I explore the historically taken-for-granted narrative on the sacred at the heart of Western culture. I offer the exploration as an alternative perspective upon a series of religiously influenced *stories* that are, in spite of being *myths*, nonetheless, often *still* presented as absolutes. Although I use the words 'story' and 'myth' interchangeably I also draw something of a loose distinction between myths and stories. Such differentiation suggests that myths generally refer to some kind of supernatural elements that elude proof. Giving credence to the existence of such elements may well have become an accepted part of the fabric of collective social life. Such stories and myths belong, nonetheless, within the sphere of subjective belief and experience, rather than the more prosaic realms of strictly verifiable accounts. Much of the motivation for the exploration I present here has been inspired by my embodiment as a member of the gender assigned the role of the profane, within the predominant Western myth about the divine, as I have sought to develop a relationship with the sacred that affirms my own, lived experiences as a woman.

Given its clear generosity towards the articulation of such attempts, I suggest that postmodernism may yet have a contribution to make to envisioning an Age of Re-Embodiments. Depending, of course, upon how we frame it. The term 'post' might, for example, be seen to infer that 'his'tory and the singular 'truth' associated with it has somehow come to an end. Now that we supposedly exist in the time period that comes 'after' we may claim to have no further need to apportion ethical responsibilities for continuing abuses or past damages done in the name of the modernity we have now, left behind.[3] Interpretations like this can be used, consciously or otherwise, as a sleight of hand, distracting attention from the remarkable continuity of any narrative story line that is told often enough. Embedded in community, culture and embodied people, even the most dysfunctional narratives have the habit of passing from generation to generation, often ad nauseam.

I, therefore, take the view that we are still living within the constructions and organizational forms known as modernity, albeit in a *hyper*modern form, rather than having succeeded in dissolving them yet.[4] Given the modern (and ongoing) obsession with the latest saleable commodity, continually throwing away yesterday's outdated invention, we might do well then to ask what we might derive from postmodernism's attempts to critique its 'predecessor'. Indeed we might use it to further the goal of accountability, choosing then, to recycle the term 'post' as part of our project to envisage an Age of Re-Embodiments. We might use it to pose the most important question in our shared planetary existence of just what does come after modernity? Will the imaginative capacity that is the product of billions of years of evolution simply lead to the apocalyptic 'post' script of the ultimate dead end? Or can we funnel our lust for 'progressive' narratives away from the merciless forward march towards the gentler, cautious, compositionist approach recommended by Latour?[5] Might we even find ways to conceive of our collective embodiment on our shared home that enable us to generate more environmentally sustainable ways of living and more social justice for more of us? This chapter seeks to do just that, by delving into some Western historical narratives in the realm of theology. Doing so, I explore the ways in which some of those old stories might be seen to be past their sell-by date and perhaps even just plain old untruthful. I also offer some seeds out of which we could grow some new stories. Stories that may, with time, and careful tending, enable us to imagine and embody new and different cultural narratives about our relationships with the sacred.

Working within the disciplines of leadership and organization, I seek to compare two contrasting theoretical approaches to the analysis of socially constructed hierarchy involving the sacred and the profane. I use input from ecofeminist theology in order to do so. In the sense that I am combining several disciplines, I take a supra disciplinary, or even anti-disciplinary approach,[6] to the creation of knowledge, merging conventional boundaries in order to highlight connections, which may otherwise remain obscured.

Describing constructive postmodernism, Charlene Spretnak[7] defines it as another additional approach (simultaneously both feminist and ecological) that may also assist us in our quest to imagine what lies beyond modernity. Within such a worldview we find a direct alternative to one that organizes its leaderships according to a hierarchical logic that generates stringent dualisms, alongside constructed forms of superiority and inferiority. Instead, life is seen as a vast interconnected web, rather than a linear chain with a male godhead. Within this web of connected *relationship* all life has an active, participant role to play. No-one element may be broken off and separated, as *all embedded being influences all other embedded being*. Any being may lead the way at differing moments. Actions and consequences emerge from the whole, rather than through any individual, though they are sometimes so complex that their mystery may defy absolutes or definitive explanations.

Perhaps crucially, within this worldview, mental abstraction cannot provide refuge from physical reality indefinitely, as the *body*, small and large, human and cosmological, is the site of all such emergence. The condition of the water and the soil has as much to contribute to our decisions about future direction as any manufactured facts and figures. Whether we choose to listen to the wisdom of the environment, or not, it seems highly probable that the 'environment' will have the last word. An ecological, constructive postmodernism recognizes that our surroundings exist as the ground of our being, regardless of any cultural constructs we may attempt to overlay atop. As Spretnak writes, 'Ecological postmodernism, then, replaces groundlessness with groundedness... This orientation acknowledges the enormous role of social construction in shaping human experience, but it also acknowledges our constitutive embeddedness in subtle bodily, ecological and cosmological processes.'[8]

I proceed then through using a framework provided by feminist spirituality and ecofeminism and influenced by such ecological postmodernisms. My intention is to reach back in time, digging beneath the surface. In the process, I hope to excavate some of the historical theological narratives and cosmological influences that such an ecofeminist approach suggests is partially responsible for a toxic inheritance of domination, still at work within modernity.[9] While both feminism and ecological perspectives include a multiplicity of voices within their broad reach, most tend towards passionate engagement in the representation and defence of diversity.[10] Many of them also draw upon the same kinds of pre-historical cognitions as those represented by many 'aboriginal' communities, recognizing the ultimately spiritual *and* highly pragmatic suggestion that humans are indivisible from the environment upon which we all depend.[11]

Sensitivity to the colonialist consumption and exploitation of Western modernity suggests that it is important to acknowledge that not all feminist spirituality or ecofeminist approaches emerge from the Non-Westernized cosmological systems, which the call for this collection suggested might help us in theorizing an Age of Re-Embodiments. They do nonetheless form part of a network of popular struggles engaged in resistance to the ecocidal tendencies of so-called civilization. From the Occupy Movement to the Zapatistas of Mexico, the Achuar of Ecuador and the Chipko movement in India,[12] such popular initiatives engage in an overall effort to tip an increasingly precarious balance in favor of social justice and environ- mental sustainability. Many such endeavors keep faith with traditions that celebrate the existence of life, death and rebirth in successive cycles, rather than with a model that sees history as a relentless forward march. Joining with many others, in anarchic webs and chaotic, disordered emergence, such endeavors suggest the noisy, rebellious perspectives of bodies, aligning with nature, ancient ancestral traditions and shamanic cultures, working together, to achieve a unity in diversity that may yet come to represent sustainable change in our current global politics. As such, they seek to exist beyond the compartmentalization of the artificial terms of core and periphery and the fragmentation of enlightenment dualism,[13] stubbornly refusing to believe that the task of imagining an Age of Re- Embodiments is futile. This task involves nothing less fundamental than finding ways to address the multitudinous species annihilation,[14] environ- mental damage[15] and social inequalities that persist as cancerous and deadly outgrowths of colonialism.[16] Though the 'logics' at the heart of modernity may be exhausted, they are far from being halted. The perspectives offered by feminist spirituality and ecofeminism are therefore presented here as a valuable theoretical contribution to the project of imagining a future that might re-embed and re-embody us within the parameters of the environment upon which we are utterly dependent, in ways involving increased co-operation and respect for all beings.

I begin by briefly outlining the first approach I analyze. Keith Grint offers this in his paper 'The Sacred in Leadership: Separation, Sacrifice and Silence'.[17] It provides an outstanding elucidation of the historically dominant, heroic model of leadership typifying the patriarchal norms and androcratic organization of 'Western Man and his externalized others', as well as its relationship with the sacred. In taking an etymological approach Grint traces the roots of Western leadership, with its division between sacred leaders and profane followers, back to Divine Kingship. Grint's work, while acknowledging the way in which societies sacralize their 'political community', is also rooted within a Judeo-Christian cosmology that is only indirectly acknowledged.[18] His essay on the sacred, is perplexing in its lack of attention to its broader theological context. Drawing from Judeo-Christian sources he nonetheless fails to acknowledge a wider truth, namely, that any *notion* of the sacred is intimately bound up with the *specific* cosmology of the society in which it is located. His analysis of the connections between the sacred and leadership reveal 'a pyramidal world ruled from the very top by a male god, with men, women, children, and finally the rest of nature in a descending dominator order.'[19] Meantime

Grint's assertion that replacing the system he outlines seems to him to be close to impossible runs the danger of reinforcing its existence as the natural order, if not the will of God.

The second approach, inspired by the work of Rosemary Radford Ruether, a feminist theologian, is employed to critique the first. Ruether provides a painstaking archaeological excavation of the successive transformations of political relationships into religious narratives and philosophical theory, as they moved through Babylonian and Hebrew societies to classical Greek Platonic thought. This results in an extensive analysis of their ongoing (and she argues, disastrous) influence within modernity.[20]

Ruether's analysis highlights a number of key elements for consideration. It reveals the ways in which the Judeo-Christian cosmology helped to disembed the divine from the everyday world. Echoing the earlier Babylonian and Hebrew cosmologies, as well as the Platonic one, with their male Gods, Judeo-Christian cosmology placed the sacred outside of nature and beyond the majority of people. Such separation, as we see in these cosmologies, served to remove goodness and value as inherent in all beings. It enabled the establishment of an instrumental approach to the 'other'. It legitimated slavery, servitude and domination, originally by the aristocratic and priestly elites in Babylonian society, over serfs and slaves and later to include the subordination of all women, Earth and animals, within the Hebrew and Platonic versions. Generating social and organizational structures fashioned after a patriarchal divinity demanding fear, conciliation and submission, based upon his inherent superiority as wholly independent creator (leader), this disembedding of creation acts as the ultimate justification for the exercise of what many feminists call power-over. Having outlined Ruether's work I conclude with some suggestions of the ways in which feminist spirituality and deep ecology are able to assist us in finding ground in which we may plant the seeds of a re-embodied and re-embedded approach to leading change.

Living by the sword

Writing in *The Sacred in Leadership: Separation, Sacrifice and Silence*, Keith Grint begins by challenging the notion that either economics or politics are in crisis. Environmental degradation, poverty and social injustice, dwindling resources, globalized asset stripping, overpopulation and multiple sites of military conflict, all are absent from analysis of an environment framed as only 'allegedly besieged… by financial crises, terrorism and political scandals'.[21] Suggesting that we need to acknowledge and engage with, rather than deny, the sacred element of leadership, Grint offers us a description that includes three very specific elements, which he argues both epitomize the sacred and actually enable leadership itself, to take place.

The bedrock of his argument is provided by an etymological approach, used to elucidate both the constitution of that which is sacred and the way in which leadership itself is contingent upon this very specific construction of the sacred. The word sacred, he tells us, 'comes from the Latin *sacer,* meaning sacred or holy or untouchable, which itself came from the Latin *sancire* – consecrate, dedicated to a religious purpose, reverenced as holy, secured against violation; to set apart'.[22] Sacred therefore denotes those who are remote from the ungodly, whom within this construction, are the followers. The sacred (leader) is intrinsically different from the profane (follower). Sacrilege, coming from a Latin compound, *sacra* and *legere,* means 'to steal holy things'; it involves the contamination of the holy, wherein the boundary of something or someone considered untouchable is breached. Hierarchy, a word composed of the Hellenic vocables – *hieros* and *arkhos,* refers to the sacral ranking of the Holy Sovereign. Hierarchy is the sacred organizational space that facilitates the leadership of God and his priesthood. The Latin term s*acerdos* means priest. Sacrifice, for its part, is derived from a Latin compound, *sacrificare* denoting 'to make holy'. It is thus demonstrated, that 'a second element of the sacred relates to the essential issue of sacrifice by those deemed closest to god – the priesthood; sacrifice is what makes something sacred – it performs leadership.'[23] To conclude, the third component of the sacred refers to the appropriate attitude or stance while in the presence of the divine, which is described as awe or silence. For Grint this silence implies two things, namely, the silencing of fear and existential anxiety as followers place their trust in God and his representatives. When all else fails there is always the suppression of opposition, 'as non-believers and heretics are silenced'.[24]

Thus Grint's account, which takes hierarchy specific to one particular cosmology, may be seen to validate one particularity as the general and therefore predominant organizational modus operandi of all leadership practices. An account that also seems to reinforce the idea that such practices must, of necessity, rather than choice, be based on separation, sacrifice and silencing. Ruether suggests that there are other cosmologies that pre-date the one referenced by Grint, but which continue nonetheless, to exert potent influences in the modern world. These then are the original cosmologies that I suggest form the foundations of the secular theology that I propose lies at the heart of Grint's thesis upon the sacred in leadership.

'When God was a woman'[25]: remembrance as an act of re-embedding and re-valuing

Employing Ruether's work we now further explore this proposal that many leadership and organizational mandates within modernity have not only emerged from but *are also still* powerfully influenced by a repressed and secular theology, embedded within a Judeo-Christian cosmology partly derived from cosmologies that preceded it. As such the mandates that originally emerged from these cosmological sources constitute part of the process of the largely unacknowledged employment of 'secularized theological concepts'[26] within leadership and organizational studies. These represent a range of collective unconscious assumptions that I suggest are problematic in a number of ways. The secularized theological concepts under discussion, privilege hierarchy and violence and involve the systematic 'othering' of that which is excluded from the elite positions in their hierarchal ordering. Within the conceptual framework provided by these theological constructs it is suggested that leadership is seen as the almost exclusive province of an all-powerful male figure standing outside the material realms – transcendent, separate and superior. A 'judgmental, dominator', such a figure has been, on a historical basis, within Western society, an integral aspect of the portrayal of God (and man in his image) as a patriarchal (white) male.[27]

I propose, therefore, that the representation of leadership provided in Grint's paper represents a particular kind of ethic, emerging from this specific theological legacy. This ethic delineates a very particular form that Eisler[28] associates with androcratic, or male-dominated leadership.[29] Leadership within such cultures is

> characterized not only by the rule of men within family and society, but also by central organization, hierarchy, class division and slavery. Frequent and organized warfare is characteristic of patriarchy and patriarchal societies are most usually ruled by men who are warriors or who control the military.[30]

I do not disagree with Grint's assessment of the level of difficulty we may encounter when attempting to displace the construction he outlines. For as Carole Christ comments, 'Once a cycle of violence is begun, it is very difficult to stop it'.[31] I do, however, address its existence as both a construction and a practical commitment within a patriarchal economic and ideological system, to particular ways of being in the world, often based upon the willingness to employ violence and sacrifice. Far from embodying the 'will of God' however, I identify this willingness as very clearly manifesting from the material, human realms, a stance that is now, more than ever, *severely* counter-productive to sustainable relationship and positive ethical value within an indivisible environmental context.[32]

In examining the aforementioned cosmologies we become able to see the ways in which God came to be *constructed* as a transcendent phenomenon; 'he' who is 'set apart' within a particular narrative context, rather than, as is so often claimed, *existing* as a representative of a singular absolute truth. We see also how the concept of sacrifice was transformed

from an experience involving birth and rebirth to one associated instead
with death and violence. And finally we witness how a divinity that
iconically balanced both life- and death-bearing aspects comes to be
replaced by a God who is to be feared more than celebrated and has violent
death as one of his most elevated aspects.

These cosmologies are, in chronological order, the Babylonian creation
myth, circa the early part of the second millennium B.C.E.; the Hebrew
creation story and Plato's Greek creation myth, *Timaeus*. *Genesis*, the
Christian myth of how God created the world was originally derived from
the Judaic Hebrew bible written in Babylon circa sixth or fifth century
B.C.E., taken in turn from the Babylonian myth *Enuma Elish*, read every
new year as an offering to renewal and continuation.[33] The Babylonian
myth has its own origins in an earlier Sumerian myth, which begins not
with God but with a primal Mother as the creator of both the Universe and
the Gods and Goddesses that populated it. Within this version the world is
not created by the word of God but born of the mother. Her sacrifice is
one based upon life-affirming creativity. She births successive generations
of divinities; the primal parents, Heaven and Earth, followed by water, air
and plants, and only finally the Gods and Goddesses, whom Ruether
suggests to us represent the ruling class of the Babylonian city states.
Throughout the nineteenth to sixteenth centuries B.C.E. these metro-
politan elites struggled to control both their own 'working classes' and
invaders from neighbouring states, migrant itinerants and the elemental
chaos that arose from droughts and floods, all of which endangered their
sovereignty.

At some point the story of the *Enuma Elish* was rewritten so that the
ancient Mother, Tiamat, would come to represent the forces of chaos,
which challenge order and control. One of the younger Gods, Ea, kills
Apsu, one of Tiamat's consorts, crowning himself king. Along with
Damkina his wife, Ea then has a son, Marduk, who in turn kills Tiamat.
Marduk then (re)creates both Heaven and Earth from Tiamat's corpse,
making humans from the blood of her other consort Kingu, whom Marduk
also kills. Given the dubious origin of humanity within this myth, it helps to
legitimate human slavery, thus freeing the Gods to pursue pleasure. In
other words, hierarchy is given a religious mandate. Leisure is acquired for
the aristocracies of temple and palace through the appropriated labour of
the serfs and slaves whose co-operation is ensured through the 'hieratic
power of ritual and law and the military power of armies and weapons'.[34]

Over the course of one generation the Babylonian dynasty replaces an
earlier matriarchal world in which women maintain autonomy, and
creation is seen to occur naturally and organically through partheno-
genetic means. In this world neither Gods nor Goddesses exist in isolation
from the world. Rather they arise out of it, as begotten and gestated beings
generated from what came before rather than magically appearing *ex*

nihilo. The sacred is embedded within and a part of, rather than separated from, human affairs.

With the rise of Marduk, however, origin is linked both with force and death, akin to Grint's conceptualization of sacrifice. Consideration of this conceptual shift offers a different lens through which to view the kinds of organizational processes discussed by Grint, which he argues are of *necessity* based upon the principle of violent sacrifice:

> Marduk extinguishes the life from Tiamat's body, reducing it to dead 'stuff' from which he then fashions the cosmos... This transition from reproductive to artisan metaphors for cosmogenesis indicates a deeper confidence in the appropriation of 'matter' by the new ruling class. Life begotten and gestated has its own autonomous principle of life. Dead matter, fashioned into artefacts, makes the cosmos the private possession of its 'creators'. Even though the new lords remember that they once gestated out of the living body of the mother, they now stand astride her dead body and take possession of it as an object of their ownership and control.[35]

In the process, divinity comes to be constructed as something separate from the earthly realms and the organic and natural processes of creation. Thus the sacred and the profane become distinct, separated entities. This, in turn, has profound implications for any realm, in this case the Earth, which comes to be seen as profane, or non-sacred. A cosmology propagating the belief that the Earth is somehow a mere resource is likely to produce a very different organizational mandate to one which relates to 'it' as a source of immanent divinity. The demotion of the 'Goddess' as a source of divinity may also be seen to pave the way for justifying gender hierarchy.

The Hebrew creation story, which I examine next, authored by those of a cognoscente priestly class well aware of the Babylonian predecessor retained some features while changing others in ways more reflective of their own 'cultic system'.[36] All conflict between Mother and Father has already been erased. God alone creates this Universe, like Marduk, out of pliable matter. Mirrored after the priestly class rather than a warrior one, at least in Genesis, God makes the Universe through the act of ritual declaration, 'Let there be...'. Like Marduk, God creates in stages, with humans coming last made in God's image and appointed to dominion over the Earth and all her other creatures – a dominion, – which, according to some scholars, establishes the roots of today's ecological crisis.[37]

Discussing the historical roots of our current ecological crisis, White also argues that, 'human ecology is deeply conditioned by our beliefs about nature and destiny – that is, by religion'.[38] Assessing the impact of the same creation story outlined by Ruether, he continues, 'The victory of

Christianity over paganism was the greatest psychic revolution in the history of our culture... Man named all the animals, thus establishing his dominance over them. God planned all of this explicitly for man's benefit and rule: no item in the physical creation had any purpose save to serve man's purpose... Christianity is the most anthropocentric religion the world has ever seen'.[39] Examining the triumph of the concept of dominion over animism he describes the removal of the need to placate the *genius loci*, or spirit of the place, which exists within cultures that see spirit as embedded in the world: 'The spirits *in* natural objects, which formerly had protected nature from man, evaporated. Man's effective monopoly on spirit in this world was confirmed and the old inhibitions to the exploitation of nature crumbled.'[40]

Some animistic culture does nonetheless survive. The kind of worldviews associated with it, as well as the emerging perspective of contemporary ecology,[41] offer alternative assemblages to disembedded and disembodied hierarchies generated by the kinds of processes outlined by White. As just one such example, David Abram describes the way in which he slowly came to understand how 'spirit' (the sacred) is embodied in Balinese culture. Observing villagers leave offerings for the household spirits, he noted that the recipients were in fact the local ants. Puzzling over how this could be equated with gifts for the ancestors, he gradually came to see that the offerings, not only fulfil a practical function of attempting to persuade the ants to stay outside the homes of the villagers, (a non-violent way of establishing a boundary?), but they also suggest a notion of the sacred based upon very different premises to the anthropomorphism and transcendence of Judeo-Christianity. Abram elaborates,

> 'Ancestor worship,' in its myriad forms, then, is ultimately another mode of attentiveness to nonhuman nature; it signifies not so much an awe or reverence for human powers, but rather a reverence for those forms that awareness takes when it is not in human form, when the familiar human embodiment dies and decays to become part of the encompassing cosmos. This cycling of the human back into the larger world ensures that the other forms of experience that we encounter – whether ants, or willow trees, or clouds – are never absolutely alien... they remain... familial.[42]

In other words, such cosmology dissolves rigid separation between heaven and earth, which when integrated become one vast interconnected and vitally embodied totality. Such cosmology may also be seen to limit superiority based on hierarchy since it seeks to recognize the connections between the non-human and the human worlds, acknowledging the interdependence of both realms. The recognition of the non-human world as ancestry seems likely, once again to generate a different kind of organizational mandate to that of dominion. In other words, it may be seen to lead to a sense of embeddedness that generates a desire to care 'for', rather than imposing the kind of control 'over' that has tended to be associated with Western cultures. Both the violence and the superiority, which are implicit in Grint's conceptualization, are absent.

Within the repressed theology of Grint's conceptualization, leadership based upon such hierarchical ordering has also, as he himself acknowledges, often been used to justify the violent sacrifice of women as well as nature. Such ordering, I suggest, is not however inevitable nor is it dependent upon inherent inferiority but is the result of specific historical processes, often rooted in religious belief systems, which have unfolded over millennia.

Writing in *Wild*, Jay Griffith documents a seven-year odyssey around the planet, to visit what is left of animistic cultures within the contemporary world. Griffith outlines in painfully exhaustive detail the ways in which similar destructive processes to those already discussed above, continue to be visited upon shamanistic peoples in South America, the Arctic, Indonesia, Australia and Outer Mongolia. Recalling various stories told to her by the Ashaninca people of the Amazon, she describes a radically different approach to that of Judeo-Christianity:

> In the punitive austerity of Genesis, the tree represents the 'evil' of sexual knowledge and the woman is sinful. Tree, woman and sex are tied into the damnation of 'the Fall', while here in the forests the stories tell the exact opposite; the lively flirtatious tree and the laughing sexy woman are the heroes of the story and she is wedded to the tree in the universal human truth of our long and necessary rapport with trees. Without them we truly fall.[43]

Taking a different perspective upon 'the Fall' I move finally to examine the repercussions of the Platonic cosmology represented by *Timaeus* written early fourth century B.C.E.. This cosmology absurd though it may now seem, was in its day regarded as scientific, until it was de-throned by the work of Copernicus and Galileo in the sixteenth and seventeenth centuries. As classically educated men, the authors of *Genesis*, while they took the Hebrew creation story as their patent, were also influenced by the milieu of the day in their interpretative (read subjective) endeavors and, Ruether suggests, 'for 1,500 years read it with the cosmology of *Timaeus* in the back of their mind'.[44]

Plato begins by establishing the dualism that has had such far-reaching influence upon Western scientific thought. Thought, which Plato sees as the pure, invisible eternal realm is, in the *Timaeus*, separated out from the material, visible domain of the body. Between the two realms the Demiurgos makes the cosmos from the four elements, a cosmos, in which the Earth is seen as the centre of the Universe. Successful souls (where the term soul is associated exclusively with masculinity in Plato's work) earn the right to ascend to their rightful place among the stars should they manage the primary task of incarnation, namely conquering the passions of the bodily state.

Creating homes (bodies) for the souls is seen as too corrupting for the Demiurgos who assigns this task to delegate Gods. Souls are schooled in the celestial sphere, prior to incarnation. Once embodied, provided they can transcend the depravity of the mortal and sinful flesh, they will be rewarded with a return to this ' 'first and better' state... as a (ruling class) male human, winning finally his return to his original disincarnate state in his star above'.[45] Failure to prevail over the unruly passions of body and emotion will result in demotion in the next return to the flesh, to a woman, with further demotion to an animal if this failure to transcend 'evil' continues. The hierarchy of mind over body, replicated in the superior/ inferior dichotomy, also applies to heaven over earth, and to male over female and animal. Although the hierarchy of rulers over workers is not detailed in this work, Plato returns to it in *The Republic*. In all three categories, which he then offers – philosopher-rulers, guardian warriors and manual workers – women are uniformly subordinate. Alongside the legitimations of hierarchy imposed by the previous cosmologies, Plato includes the profoundly important element of reinforcing the relegation of the earth and adding the body, alongside women, workers and beasts, to the status of the 'fallen'.[46]

The process of conceptualizing divinity so that it is seen to exist only in specific places, people and elements, (such as heaven, masculinity and mind) inevitably leads to the establishment of hierarchy. Such segregation functions to generate and maintain the superior value of those qualities and/or people deemed sacred. This in turn confers an inferior status upon those states viewed as profane. The subjugation of earth, female, animal, body and emotions is thus legitimated, and paves the way for all of them to be treated as the superior (divine) leadership deems appropriate. The earth becomes a resource, or capital, and neither 'the body', nor 'its' feelings, need be listened to or given equal status to that which is viewed as sacred – in this case heaven and mind. Divine (and frequently masculine) leadership has a mandate to 'silence' lesser men, women, children and others labelled as profane and therefore as inherently more disposable beings. Ruether suggests that alongside, law and philosophy, theology becomes the "'master narrative' or 'logic of domination' that defines the normative human in terms of this male ruling group".[47]

It is obviously important to acknowledge the existence and role of various Christian liberation theologies around the world, the stewardship model and the great diversity of Christian thinking, including the role of heresy. I would nonetheless argue that these alternatives have not constituted a systematic or widely accepted approach that have been able to replace the dominant discourse. It is this dominant discourse that I attempt to address here. Rather this work takes the view that some strands of Judeo-Christianity have been harnessed on a long-term basis to sacrilize the political relationships of domination and that these need to be addressed. In this sense, I speak to the history of some of the cultural, discursive or narrative and mythological elements associated with the Judeo-Christian religions that have been used to maintain hierarchical and oppressive organizational and leadership structures. While I do not intend wholesale condemnation of Christianity, such elements, which have included a messianic religious ideology and militarist imperialism, will require both uncoupling and re-evaluation if Christianity as a whole is to provide an ethical framework that serves constructive change, environmental sustainability and social justice. This work then, is a critique of traditional and conservative, rather than of liberal or radical approaches to Judeo-Christianity, as it argues that they are still one of the most powerful influences upon our understanding of the sacred within Western culture.

In distinguishing between different kinds of influence, some feminists have described *power-over* as the fundamental control structure associated with such male dominated, androcratic cultures. Based upon domination, power-over functions through the subordination of those regarded as inferior in interconnected and inter-related patterns of gender, race, class, age and ethnicity. Such relationships may be seen to function on different but nonetheless related levels from the ideological and cultural narrative level to the socio-economic level.[48] As Sullivan notes: 'global ordering is made not only at macro and formal institutional levels, but also through more quotidian forms of power, constituted around gender [and sexuality] but also intersecting with hierarchies of race and class'.[49] Within such embedded systems of power-over, violence, ownership, superiority and control become so pervasive that they are all but invisible, representing taken-for-granted organizational norms. Those belonging to the inferior, profane classes are viewed as lacking the characteristics that would enable them to lead. This provides the justification for their exclusion from forums of power and decision-making and for their ordering as 'other.'

Restoration of value through a transformation of the sacred

A *mythos*, or mythology is, as has been suggested in the previous section, a socially constructed phenomenon, consisting of many interwoven symbols, narratives and ritual enactments. In spite of much of modern leadership's insistence that it represents the ultimate in rationality, the *mythos* at the heart of such claims is far more than just a cerebral system of values. Symbols influence both the conscious and the unconscious worlds, as repetitive ritual enactments act to define what is and is not of value. This is knowledge that also becomes deeply embodied.[50] As Z. E. Budapest writes, 'that which is not celebrated, that which is not ritualized, goes unnoticed, and in the long run... will be devalued.'[51] Restoration of value, in turn, requires the ground of a new mythos as the seedbed in which a new ethos can be grown. We turn now to feminist spirituality and deep ecology in search of new ground.

As has already been acknowledged here, both feminist spirituality and ecofeminism embrace a wide diversity. Although many of the scholars representing these viewpoints are critical of patriarchal religion, not all employ the symbolism and imagery associated with the Goddess in all of their work.[52] Among those that do there are still areas of disagreement as to the precise meanings that they attach to this word. Some believe categorically that archaeology provides extensive evidence of cultures during the Paleolithic and Neolithic eras that incorporated the Goddess as the central deity and ordered reality very differently from what was to come after.[53] Others problematize what they see as a mythology of a 'prepatriarchal paradise and the fall into patriarchy'.[54] In a world with an expanding human population and increasingly sophisticated technology, as well as many environmental and cultural wounds that need tending, this chapter acknowledges that there can be no return to the past. It chooses, nonetheless, to explore the metaphor of the Goddess, as to do so provides a useful lens (as one among many) with which to examine the damaging assumptions within the Judeo-Christian cosmology that the chapter argues have influenced the construction of leadership and organization in Western culture. The Goddess also holds the potential to contribute to what Irigaray describes as an

> attempt to consider the possibility of a maternal genealogy and the symbolic and institutional forms it may take... not to be a reversal, the simple replacement of patriarchy with matriarchy, but rather the coexistence of two genealogies.[55]

The God of the Judeo-Christian tradition is not only male but is also transcendent, separate and effectively disembodied. While it is perhaps important to acknowledge the incarnate aspect of God in the figure of Christ, as well as some biblical descriptions of the creator as female, my own personal (and therefore subjective) reading of Christianity, is that it is predominantly male-dominated and body negative, particularly, in relation to women as the origin of worldly sin. Often represented as heavenly light, God symbolically illuminates the darkness of the profane, earthly and sinful world. The Goddess on the other hand, is female, earth, nature and body, in ways that include the wisdom of emotion, rather than the need for its suppression.[56] She balances both light and dark, through associations such as sun and moon, night and day, life and death and summer and winter.[57] The earth as immanent divinity is not an abstraction that we experience as separate from the world, but is, as our home and our mother, the source of all life. Awe within such an immanent configuration may then be reviewed to mean both respect and reverence for that which brings forth and sustains life: in women, or the earth, or in the multitude of daily acts that both genders perform in order to support life. Awe can be relocated, symbolically and literally (not as fear), as the celebration and appreciation of beauty and uniqueness and in the nurturing of life.

As Carole Christ points out, the fact that the Earth (beyond the artificial dismemberments of colonial and neo-colonial geo-politics) encompasses all nations and peoples means that no one group is enabled to claim exclusive access.[58] Indeed if the divine is seen as immanent within all of the material, anything we do to damage the Earth, or the body, also harms the divine. Bodies are not seen as inferior, corrupt or in need of transcendence. They and all their processes and feelings, including sexuality, birth and even their death, are embraced as a source of wisdom.[59] Allowed to be, they do not need to be kept under control. This means that spiritual expression requires care and repair, rather than brutal or martyred sacrifice. The Earth and 'all her relations', including nature, women, children and the colonized are re-sacralized within this framework. Value is restored, requiring a different ethos to be expressed. Relationships between equal subjects call for different behaviors.[60] Salvation becomes a profoundly embedded, practical activity, grounded in the here and now, calling for the demonstration of respect for difference and diversity.

As a symbolic representation of the web of life, the Goddess also stands for unity – with all of life. Recognizing that we all come from one source there is no basis for the superior/inferior dichotomy, either among humans or between humans and non-humans. This does not mean that conflict is absent, but when difference is perceived as part of immanence, it is recognized that violence that damages the integrity of the one also injures the many.[61] We cannot harm another without also harming self. This does not of course mean that harm does not occur, sometimes unavoidably, but there is an inherent recognition that consequences reverberate beyond the act itself. Among some indigenous people this is stated in the concept that our behavior will affect the next seven generations. Given these insights, healing and helping or harming become the pragmatic alternatives to good and evil, as embodied and daily choices. Constructive choice within this ethos arises from the feeling of being connected, wherein the individual is inextricably linked to community.

It is a viewpoint very similar to that expressed in the following quote, normally attributed to Chief Seattle.

Humankind has not woven the web of life,
We are but one thread within it,
Whatever we do to the web,
We do to ourselves,
All things are bound together,
All things connect.[62]

This philosophy is also expressed as deep ecology.[63] In many ways the philosophy at the heart of feminist spirituality, ecofeminism, deep ecology and much of the mythoi expressed by many of the indigenous peoples, who have been colonized and exterminated has a great deal in common. Far from being savage, or uncivilized, indigenous peoples often express a sensibility that has much to offer to the world. Writing in *God is Red*, Vine Deloria comments, 'Christianity has traditionally appeared to place its major emphasis on creation as a specific event while the Indian tribal religions could be said to consider creation as an ecosystem present in a definable place.'[64] Similar connections are drawn together in Charlene Spretnak's work, *States of Grace*, which examines a variety of wisdom traditions. Part of her commentary is upon the difference in attitude between what she describes as primal traditional cultures and modernity. Ultimately, Spretnak outlines the variety of different, but interconnected, ways in which native culture, ecofeminists, deep ecologists and bioregionalists all identify our collective need to develop our awareness of our place in the greater web. Each of these traditions recognizes our inherent dependence upon a web of interrelatedness; a web that is seen as worthy of respect in its own right and not merely as a cipher for human activity.[65]

Writing specifically about deep ecology, Patrick Curry highlights the fact that the energy and motivations for the implementation of change often emerge from sources that are far from rational or secular. He argues that emotion, if not in fact love, has a crucial role to play in changing our attitude to the environment, identifying both religion and spirituality as sources of support for constructive change.[66]

The distinction between the spiritual and the material is framed as the fatal flaw in the cosmological systems that I have critiqued here. Curry is very clear about precisely what sort of spiritual and religious qualities he views as being helpful to the world at this time, suggesting a need for inclusive pluralism; a sense of place (local); the recognition of ultimate mystery while maintaining involvement in the material; the celebration of the everyday; and a social rather than an individualistic bias: 'What is needed is to encourage and strengthen people's awareness and appreciation – which already exists, although it is rarely articulated – of the Earth and all its life as sacred: not an abstract Life, but one that is embodied and embedded in specific relationships, communities and places.'[67]

Re-embodied leadership: ancient ancestral wisdoms for an age of re-embodiments

Espousing a similar outlook, the leaders within the business community in which I recently conducted empirical research for my PhD thesis, comment, 'It is no longer a matter of being alternative. It is a matter of there being no alternative'.[68] There are alternatives, but should we choose to exercise them, we will all pay the price in increasing environmental degradation, warfare, suffering and, potentially, irretrievable loss.[69] For, as Curry observes, 'life... is entirely relational... the entities related are constituted by those relations – and reflexive, so that it is impossible to stand outside and observe or manipulate it, [life] either as a whole, or in part, without... being affected by it.'[70]

The illusion that it is possible to separate the part from the whole all too easily results in destruction to relationships and the material base they are dependent upon. The alternative to the false logic of dismemberment is to engage with the seemingly overwhelming task of attempting to manifest new ways of being that also seek to encompass the best of the past, in an Age of Re-Embodiments. This suggests revision of many central aspects of the ideology and practices that have organized our world and our ways of being for hundreds of years. This chapter has examined a model of leadership that has valorized just this kind of artificial separation. Casting a transcendent and superior male God, in the main role in its dramatic narrative, the dominant model has often also functioned as a device that removes the heroic individual leader from connection and relationship, from consequence and belonging.

Life in an Age of Re-Embodiments from an eco-feminist point of view does not suggest replacing such constructions with another one-size-fits-all model. On the contrary, ecofeminism is deeply committed to pluralism and diversity. Many religious traditions are currently in the process of revision[71] and, of course, there will always be those of a more secular persuasion that make deep and valuable contributions to our contemporary problems. Attempting to avoid a situation in which any approach becomes 'written in stone', Carole Christ suggests nine ethical principles that she calls touchstones. An ethical framework that resonates with some of the indigenous wisdoms already discussed, they include the following: to care for life; to walk in beauty; to trust bodily knowledge; to be honest about conflict and suffering; to take only what we need; to consider consequences for seven generations; to take life only after great consideration; to practice generosity; and to work to repair the web of life.[72] I propose that incorporating at least some of them holds the potential to move beyond an ethos based upon domination, towards re-embodied (as opposed to what I have described here as disembodied) ways of being.

I also suggest that re-imagining good leadership as one aspect of a versatile Age of Re-Embodiments is grounded among other things, in the willingness to question and challenge the status quo. The taken-for-granted ways of doing business are killing us – quite literally, threatening us with the ultimate and permanent disembodiment in the form of species extinction.[73] I propose that, 'leadership should be aimed at helping to free people from oppressive structures, practices and habits encountered in societies and institutions, as well as within the shady recesses of ourselves'.[74] I argue that an abhorrent ethos based upon the dualism of superiority and inferiority has created a worldwide crisis that has emerged from the violence associated with a disembodied philosophical position. This needs to be replaced by an ethic based upon the willingness to heal the damage done by such violence. An ethical framework built upon connection rather than separation suggests that such a task, along with the leadership implied by it, is a shared one, in which we all have a role to play.

Such an ethic values the body and feelings, as well as the mind and rationality, leading to a renewed respect for them. Leadership, rooted in both compassion and understanding of life as emergent process, located within us all, also works to nurture and honor all aspects of being human, without elevating certain qualities as superior. Embracing the under-standing that there is no inherent separation between the spiritual and the material, such leadership accepts the diversity of ways humans choose to live their lives as a matter of personal choice. This kind of approach is therefore, equally inclusive of those who prefer a secular ethic, while also accepting that other people are likely to continue to use words like sacred and divine in order to describe and share their experiences and under-standing of the ineffable. Nonetheless, it recognizes that the ways in which we construct the sacred are of pivotal importance in how we relate to leadership, following, organization and the many different ways in which we all go about making our contribution to the world.

At a time in our history as a species when we may rapidly be reaching a crossroads, the kind of organization we mandate is of crucial and fundamental importance. Leadership that is unable to perceive immanent, inherent value in that which surrounds it cannot be part of a tentatively emerging Age of Re-Embodiments. If salvation is conceptualized as existing outside the realm of the material (where this means body and Earth), this idea may discourage care for that which is, after all, only finite. If the act of creation itself and the notions of salvation that lie at the heart of Judeo-Christian cultures are disembedded from the material realm, this, it seems, has profound implications – for all life. If the interests of the majority (rather than an elite minority) are to be served, we have a pressing need of leadership focused on a combination of social justice and environmental sustainability. Encouraging a new kind of awareness based upon a synthesis of these concerns may then lead us towards new kinds of social organi-zation and fulfillment. It seems to me that these concerns are indeed being taken up at this time by popular struggles around the globe that may yet become the ancestors of a life-affirmingly rooted Age of Re-Embodiments. Many of these struggles draw upon their love of the Earth. The embodiment of the kind of awe associated with love, as opposed to fear, may yet bestow us with a sense of belonging that motivates a different kind of predominant attitude to our world. Belonging, may yet lead us towards something both ancient and freshly imagined.

Notes

1   Y. S. Lincoln, 'What a Long Strange Trip It's Been... Twenty-Five Years of Qualitative and New Paradigm Research', *Qualitative Inquiry*, 2010, vol. 16, 3–9.

2   A. Coffey, *The Ethnographic Self: Fieldwork and the Representation of Identity*, London: Sage, 1999, p. 11.

3   E. Said, *Culture and Imperialism*, London: Chatto and Windus Ltd, 1993.

4   A. Giddens, *The Consequences of Modernity*, Cambridge: Polity Press, 1991.

5   B. Latour, 'An attempt at a "Compositionist Manifesto"', *New Literary History*, 2010, 41: 471–90.

6   K. Ferguson, 'On Bringing More Theory, More Voices, and More Politics to the Study of Organization' *Organization*, 1994, vol. 1, 81–99.

7   C. Spretnak, *The Resurgence of the Real: Body, Nature and Place in a HyperModern World*, New York: Routledge, 1999.

8   Ibid., p. 72–3.

9   R. R. Ruether, *Gaia & God: An Ecofeminist Theology of Earth Healing*, New York: HarperCollins Publishers Ltd, 1992.

10   *Integrating Ecofeminism, Globalization and World Religions*, Maryland: Rowman & Littlefield Publishers, Inc., 2005.

11   C. Bullis and H. Glaser, 'Bureaucratic Discourse and the Goddess: Towards an Ecofeminist Critique and Rearticulation', *Journal of Organizational Change Management*, 1992, vol. 5(2), 50–60.

12   B. Swimme and T. Berry, *The Universe Story: From the Primordial Flaring Forth to the Ecozoic Era*, New York: HarperOne, 1992.

13   R. Tarnas, *The Passion of the Western Mind: Understanding the Ideas that Have Shaped our World View*, New York: Ballantine, 1991.

14   P. Curry, *Environmental Ethics: An Introduction*, Cambridge: Polity Press, 2011.

15   P. Higgins, *Eradicating Ecocide*, London: Shepheard Walwyn Publishers Ltd, 2010.

16   B. Banerjee, 'Necrocapitalism', *Organization Studies*, 2008, num. 29, 1541–63.

17   K. Grint, op.cit., 2010.

18   R. Girard, *The Scapegoat*, Baltimore, MD: Johns Hopkins University Press, 1982.

19   R. Eisler, The *Chalice & The Blade*, San Francisco, CA: HarperCollins, 1987, p. 162.

20   Ruether, op. cit., 1992, p.10.

21   Grint, op. cit., p. 89.

22   Ibid., p. 91.

23   Ibid., p. 91.

24   Ibid., p. 92.

25   M. Stone, *When God Was A Woman*, UK: Virago Ltd, 1976.
26   B. M. Sorensen, S. Spoelestra, H. Hopfl and S. Critchley, 'Theology and Organization', *Organization*, 2012, vol. 19, 267–79.
27   C. Christ, *Rebirth of the Goddess*, New York: Routledge, 1997, p. 23.
28   Eisler, op. cit.
29   J. Ford, 'Discourses of Leadership: Gender, Identity and Contradiction in a UK Public Sector Organization', *Leadership*, 2006, num. 2, 77–99.
30   Christ, op. cit., p. 59.
31   Ibid., p. 177.
32   Curry, op. cit.
33   R. R. Ruether, 1992. See also I Mendelsohn, *Religions of the Ancient Near East: Sumero-Akkadian Religious Texts and Ugaritic Epics*, New York: Liberal Arts Press, 1955 and E. W. K. Mould, *Essentials of Bible History*, New York: Ronald Press, 1951.
34   Ruether, op. cit., 1992, p. 19.
35   Ibid., p. 18.
36   Ibid., p. 16.
37   L. White, The Historic Roots of Our Ecologic Crisis', *Science*, 1967, vol. 155, 1203–7.
38   Ibid., p. 1205.
39   Ibid., p. 1205.
40   Ibid., p. 1205.
41   D. Abram, *The Spell of the Sensuous*, New York: Vintage Books, 1996, p. 276.
42   Ibid., p. 16.
43   J. Griffith, *Wild. An Elemental Journey*, London: Penguin, 2006, p. 64.
44   Ruether, op. cit., 1992, p. 16.
45   Ibid., p. 24.
46   Ibid., p. 26.
47   Ruether, op. cit., 2005, p. 92.
48   Ibid., p. 91.
49   S. Sullivan, 'Supposing Truth Is a Woman?', *International Feminist Journal of Politics*, 2011, vol. 13, p. 219.
50   E. Bell and S. Taylor, *The Promise of Re-enchantment: Organizational Culture and the Spirituality at Work Movement*, in D. Boje, B. Burnes and J. Hassard (eds), *The Routledge Companion to Organization Change*, London: Routledge, 2011.
51   Z. E. Budapest, *The Grandmother of Time*, San Francisco, CA: Harper and Row, 1989, p. xxi.
52   R. R. Ruether, op. cit., 1992.
53   R. Eisler, 1987.
54   R. R. Ruether, op. cit., 1992, p. 8.
55   M. Whitford, *The Irigaray Reader*, Oxford: Blackwell Publishers Ltd, 1991, p. 23.
56   Starhawk, *Dreaming The Dark: Magic, Sex and Politics*, Boston, MA: Beacon Press, 1982.
57   C. Christ, op. cit.
58   Ibid.
59   Ibid.
60   P. Curry, op. cit., p. 3.
61   C. Christ, op. cit.

62   Translated by H. Smith in 1854 and reproduced by F. J. Grant, *History of Seattle*, Washington. New York: American Publishing and Engraving Co., 1891. Recent scholars have queried its authenticity. It has nonetheless become a powerful and widely reproduced statement of an ecological approach to humanity's place in the grand scheme of things.

63   A. Naess and G. Sessions, *Basic Principles of Deep Ecology*, 1984. Available online at: http://theanarchistlibrary.org/library/arne-naess-and-george-sessions-basic-principles-of-deep-ecology (accessed 26 March 2013). Retrieved on 7th October 2011 (from the anarchistlibrary.org).

64   V. Deloria Jnr, *God is Red: A Native View of Religion*, New York: Putnam, 1973.

65   C. Spretnak, *States of Grace*, New York: HarperCollins Publishers, 1991.

66   P. Curry, op. cit., p. 144.

67   Ibid, p. 143.

68   S. Darling Khan and Y. Darling Khan, *Movement Medicine: How to Awaken, Dance and Live Your Dreams*, London: Hay House UK Ltd, 2009, p. xiiii.

69   P. Higgins, op. cit.

70   P. Curry, op. cit., p. 8.

71   R. R. Ruether, op. cit., 2005.

72   C. Christ, op. cit.

73   C. Bullis and H. Glaser, op. cit.

74   A. Sinclair, *Leadership for the Disillusioned: Moving beyond Myths and Heroes to Leading that Liberates*, Australia: Allen and Unwin, 2007, p. 322 (Kindle version).

Chapter 5

From enlightenment to enchantment

Changing the question

Patrick Curry

My starting-point is contemporary crisis: most fundamentally and danger-ously, ecocrisis. However, the chief features of that crisis – climate change and biodiversity crash, including habitat destruction or degradation and mass species' extinctions – are more-or-less direct consequences of industrial capitalism and human overpopulation. Its causes being anthro-pogenic, it would therefore be more accurate to describe ecocrisis as ecocide. Furthermore, having no other home or mode of existence, humanity is also destroying the basis of its own integrity, viability and ultimately existence: hence ecocrisis is also a human crisis.[1]

With this situation in mind, I want to consider Kant's response in 1784 to the question, 'What is Enlightenment?'[2] and Foucault's reprise exactly two hundred years later,[3] as well as how Matthew Taylor recently took up the same question (even more briefly).[4] The point is to see, first, how their answers stand up now; second, whether those answers now point towards a better response; and third, whether indeed a better question emerges from that response. In doing so, I'm not under any illusion that philosophy leads the way, so to speak. Nonetheless, philosophies or worldviews or meta-physics do, it seems to me, play a role in what happens beyond that of mere epiphenomena or ideological window-dressing; so it is defensible, and in some ways and contexts helpful, to question them. I also have no interest in psychobiography; what matters here are discourses, especially as others who may not ever have heard of their authors influentially take them up. But I want to add that 'discourse' refers not to putatively abstract theory (as if such a thing were possible) but to practices, including theoretical practices, which are therefore necessarily both embodied and embedded.

Kant's enlightenment

Kant's essay targets humanity's immaturity, which he attributes to 'the inability to use one's understanding without guidance from another.'[5] It needs no hindsight to be struck by his negative construal of seeking guidance for how to know and act; in blaming 'laziness and cowardice', he leaves no room for humility, for example.[6] The corresponding positive virtue for Kant is, famously, exercising 'the freedom to use reason publicly in all matters.'[7] Hence 'the motto of enlightenment', '*Saper Aude!*': dare to know! Only that will result in more true knowledge and fewer errors: the enlightenment he identifies with 'human nature, whose essential destiny lies precisely in such progress.'[8] Then there is the politics of the Emperor Frederick II's position that Kant extolls in closing, to the disadvantage, significantly, of republicanism: 'Argue as much as you want and about what you want, but obey!'[9] That is, obey the chief political authority.

Such a crudely schematic appraisal yields only the bare bones of the Enlightenment, but it is enough to confirm three things. (1) The impulse given it by its most influential voice was profoundly masculinist as well as rationalist, downgrading emotion and the body and, by clear implication, the feminine and Earthy. (2) In identifying the progress resulting from reason as humanity's destiny, it was teleological as well as universalist, but in strictly anthopocentric terms. And (3) that the political dimension of that universalism was specified as narrowly discursive. Another point, perhaps less remarked, is Kant's almost monadic individualism, in which any 'external' guidance is to be rejected.

So far, so well-known; but let me remind you that we are interested here not in Kant but in the extent to which these values and views subsequently became part of dominant disembodying discourses. The ground was already well-prepared, of course, by Cartesian rationalism, dualism and scientism, itself drawing upon Christian and Platonic theology. So to keep even the Enlightenment from distracting us, I will borrow the approach of the late Stephen Toulmin and encapsulate it as the moment when the project of modernity, which predated and has arguably outlived the Enlightenment, achieved a measure of self-consciousness. Of course, to speak clearly of 'modernity' is hardly less general or demanding, but Toulmin's excellent account makes it easier to do so.[10]

Foucault's enlightenment

Foucault suggests as much, describing Kant as setting out 'the attitude of modernity', and modernity itself 'rather as an attitude than as a period of history.'[11] I agree, although we may link the two by supposing that in all times and places, most attitudes are present in some form or another, but in certain historical epochs and locations, some attitudes will be encouraged and become dominant while others are discouraged and thus less so.

Foucault points out that Kant links 'the universal, the free, and the public uses of reason',[12] which combination then becomes the criterion for what constitutes reason. This raises the question, what becomes of reason if it turns out, not just empirically but in principle, that it is never and cannot be universal (identical everywhere and always), free (unconstrained) and even public (unaffected by personal and/or power–political considerations)? The result would not qualify as reason for Kantians and other rationalists, but the rest of us need not regret losing the conception of rationality that Bernard Williams has criticised as rationalistic.[13] It was largely in order to defend such a conception and distinguish legitimate from illegitimate reason, Foucault suggests, that Kant embarked on his Critiques; one result is that 'the Enlightenment is the age of critique'.[14] As Bruno Latour remarked, 'anyone who has never been obsessed by the distinction between rationality and obscurantism, between false ideology and true science, has never been modern'.[15] Here, as so often, religious roots are apparent, this time in the iconoclastic 'critique' of idolatry.

Foucault also notes shrewdly that since progressive change can only be self-initiated, it has to be supposed that 'man' can escape immaturity 'only by a change that he will bring about in himself'.[16] Thus for Baudelaire, whom Foucault sees as an exemplar, the modern is 'the man who tries to invent himself'.[17] Thus both individually and collectively as 'man', the Enlightenment's 'autonomous subject' – which includes its Romantic version, in this and other fundamental respects – starts to resemble a flight from relationship (both dependence and interdependence), which threatens to terminate in outright and, ironically, extremely childish solipsism. But Foucault substantially agrees with Kant, although he replaces the latter's search for the necessary epistemological limits to knowledge with the historical ('archaeological', 'genealogical') study of 'whatever is singular, contingent and the product of arbitrary constraints'.[18] Such 'permanent critique of ourselves' (an echo of Trotskyite, and later Maoist, 'permanent revolution') is nonetheless in the service of 'a permanent creation of ourselves in our autonomy'.[19]

Finally, Foucault rejects what he calls 'the "blackmail" of the Enlightenment', namely the demand to either accept or reject it *en tout*, in a simplistic and authoritarian way.[20] Rather, 'We must try to proceed with the analysis of ourselves' – which, for Foucault as for Kant, is the only kind that matters – 'as beings who are historically determined, to a certain extent, by the Enlightenment.'[21] And what is at stake in such an analysis is the question, 'How can the growth of capabilities' – that is, 'our' capabilities – be disconnected from the intensification of power relations?'[22]

Just to show that such questions retain some cultural vigour, I will also mention a more recent exposition, Matthew Taylor's lecture in 2010 for the Royal Society of Arts on '21st-century Enlightenment'.[23] Drawing on Tzvetan Todorov's recent *In Defense of the Enlightenment* (anatomised by John Gray, predictably but no less accurately for that, as childishly fundamentalist),[24] Taylor selects three Enlightenment values for modernisation: autonomy, universalism and human ends. Autonomy, he says, needs to be supplemented by an awareness of our social and natural dependencies. Universal human rights too depend on widening and deepening our capacity for empathy. Lastly, the progress of 'human ends' is imperilled by the steady erosion of reasoning about ends, including ethics, by a bureaucratic emphasis on the rationality of rules.[25] (Note, however, that the rights and the ends remain purely human.)

Re-evaluating the enlightenment

Max Weber, who worried about 'formal' reason about means inexorably replacing 'substantive' reason, voiced Taylor's last concern much earlier.[26] So this might be the place to remind you of Weber's famous definition of modernity: 'The fate of our times is characterised by rationalisation and intellectualisation and, above all, by the "disenchantment of the world"'.[27] (One could adduce Marx, among others, in related vein; but don't forget that Weber had the advantage of knowing, and taking seriously, the work of both Marx and Nietzsche.) And what is the contrast class, as it were? It is 'The unity of the primitive image of the world, in which everything was concrete magic, [which] has tended to split into rational cognition and mastery of nature, on the one hand, and into 'mystic' experiences, on the other'.[28]

As Foucault himself admitted towards the end of his life, it was unfortunate that he had not read the work of Frankfurt School (which grew directly out of Weber's) earlier on in his own.[29] If he had, he would have encountered a critique of modernity arguably more radical than his; for despite Foucault's reputation, he never renounced reason but rather redefined power to include and arguably even thereby entrench it. In which case, the argument between him and Habermas was domestic, concerning the best way to deliver the goals of the Enlightenment-in-modernity; there was no critique of those ends as such. But I will resist the temptation to stray further into intellectual history and return to the question, where are we now?

In response, I want to begin by evaluating those ends themselves in the light of the crisis I mentioned at the outset. It will help us to do so if we realise the extent to which contemporary modernity entails not the failure but the fulfilment, however perverse, of the Enlightenment programme. And by 'modernity', to clarify sufficiently to proceed, I mean the triple engine of capital, techno-science and the nation-state, driving and being driven by an ideology of progress for man (now politely redefined as 'humanity') through 'the rational mastery of nature'.[30] This dynamic and ambition have certainly survived, and survived into, postmodernity, even if most of their political and popular legitimacy has not.

I must also acknowledge that this overview is limited to the developed (or overdeveloped) world, and even here, the obligation to generalise means that many of its citizens are more objects of the trends I identify than participating subjects. However, globalisation means that careful extension to the developing world too may now be legitimate.

Most fundamentally, then, we see the triumph of the human subject – in practice, still largely if not overwhelmingly male – in pursuit of overwhelmingly anthropocentric ends, and still considering itself autonomous – which is to say, putatively disembodied and disembedded, socially as well as ecologically, with any significant others always in danger of being treated contemptuously if not brutally. (No 'external guidance' here!) We see reason – now realised as bureaucratic, economistic and scientistic rationalisation – and universalism, in its most successful form: the market logic of commodity capitalism. (These are now applying their demonic ingenuity to the commodification of ecocrisis itself: a market price for 'ecosystem services', carbon trading, 'green' technologies, etc.).[31] We see the continuing hypervaluation of 'progress', similarly defined, such that any resulting problems are treated as only susceptible of solution by more progress. We see the continuing rejection of any limits, whether in human or nonhuman nature, in principle. And many if not most of us are quite free to argue as much as we want and about what we want – blogs being an obvious instance – but most of us obey... Having considered the alternatives, how we obey.

All these things without exception have been ably and thoroughly criticised, and there is no need for me to rehearse that here. I also don't mean to suggest that there is no significant resistance whatsoever. All I want to do is propose that in the tiny corner or, more optimistically, dimension of modernity broadly called philosophy, given the 'success' of modernity on such a murderous as well as suicidal scale, it is time to stop trying to refine the ideals of the Enlightenment and replace them with something else; above all, perhaps, something more humane.

In trying to do so, we cannot begin *ab initio*. There is no scratch, bottom line, or Year Zero from which to start;[32] so we will, as Foucault says, be influenced by what we are trying to replace.[33] That does not commit us to either unthinking repetition or point-by-point opposition, however. Nor should we be afraid of correcting pathological imbalance by trying to move in another direction; that too would be to succumb to intellectual blackmail. In any case, I am not raising a placard that says, 'Smash the Enlightenment!' but suggesting an alternative that would, in practice, be a supplement. And I agree with Foucault that it is 'necessary to make the future formation of a 'we' possible': a 'we' that is not previous to questioning but its result'.[34] Our starting-point can only be our historically and socially situated selves, but that does not entirely dictate their future formations.

From enlightenment to enchantment

Given all this, let us enquire what the minimum *desiderata* are for a viable alternative, one that involves a return to, recognition and revaluation of and even reverence for what makes life possible and, arguably, worth living. Shall we start with the Earth, and the earthy? And, integrally, bodies and bodiment?[35] And equally, sex-gendered difference – especially, given its ongoing suppression and repression, the feminine? And last for now but certainly not least, relations, relationships and the relational? Which is also to say, both the ecological and the ethical.

In order to help refocus our attention on, and indeed desire for, such matters, I think it might help now to ask a different question: not, what is Enlightenment? but, what is enchantment? For insofar as Weber's insight about the terminus of enlightened modernity is correct, enchantment is just what it has occluded, suppressed and attempted to destroy. As Zygmunt Bauman put it, 'The war against mystery and magic was for modernity the war of liberation leading to the declaration of reason's independence.... To win the stakes, to win all of them and to win them for good, the world had to be de-spiritualised, de-animated: denied the capacity of subject.'[36] In searching for a radical alternative, therefore, enchantment would seem to be a logical place to start.

In a nutshell, enchantment is about *wonder* as modernity is about *will*; and what is needed is not a more efficient or refined will, but will qualified, contextualised and hopefully guided, even restrained, by something else. By the same token, enchantment is unbiddable; it can be invited but definitively not commanded. Hence the ancient understanding of *faërie*, which precisely coincides with the wild (*faërie* is nothing if not ecological) as 'the ancient universe that prevails here on Earth wherever human beings are not in control'.[37] It is not anthropocentric, let alone Promethean or Faustian. Nor androcentric; even in classical myth, arguably already decadent,[38] its strongest exemplars are Aphrodite and the sexually ambiguous Hermes.

I am also taking a hint from Weber here; for 'concrete magic' is, if you remember, just what is lost in the process of modernist disenchantment.[39] The term is apt. The *sine qua non* of enchantment is that it is an experience and world that is both 'spiritual' ('magic') *and* 'material' ('concrete'). In enchantment, both those supposedly foundational distinctions that we have had drilled into us in recent centuries – between experience (epistemology) and world (ontology), and between spirit or mind and matter – appear in all their sectarian contingency. So too does the impassable gulf between knowing subject and known (but according to Kant, ultimately unknowable) object. And what replaces the object is another subject – in enchantment, 'an object is an incompletely realised subject' – and a world/experience that is entirely relational rather than causal: 'nothing has happened but everything has changed'.[40] For the same reason, enchantment is entirely participatory, and where there is apparently no participation (such as in external observation or control, 'objective' assessment, etc.), it too is absent.

'Here on Earth': enchantment is not some off-planet heaven, or hell. It is transcendence in immanence, in which bodiment and embeddedness are absolutely integral: the place where we started, to coin a phrase, but known for the first time. Simultaneously 'concrete' – *this* place, *this* person, *this* music, *this* food – and 'magic': ineffably spiritual, unplumbably mysterious. This Earth itself, for example, in all its complex and subtle particulars. And ourselves, when we are enchanted.[41]

Now it should not be surprising, given its energy and ingenuity, that the modernist hybrid of capital, technoscience and the state has already embarked on colonising and enclosing the very things that make enchantment possible, especially the Earth and the living 'material' body, and on mastering and managing enchantment itself. In addition to the colonisation and commodification of 'ecosystem services' that I have already mentioned, there are the ultra-sophisticated methodologies (well-funded, significantly) of modern bioscience. And what else is the multi-billion-pound industry of advertising and public relations, to say nothing of its close and almost equally profitable relative, electronic pornography? If I am right, however, these enterprises will, or rather must, fail – and that, to the very extent that they apparently succeed.

Why? Because if enchantment is, as I maintain, inalienably wild and unbiddable, then what is being successfully produced and managed in order to target consumers and generate profit is something else: a simulacrum of enchantment. Indeed, if enchantment cannot be captured alive, this simulacrum is its externally animated corpse. In any case, to mark the difference, I call it 'glamour'. Glamour bears much the same relationship to enchantment as pornography does to erotic love – not coincidentally, a principal site of enchantment.

Concerning the body and materiality, I have emphasised their centrality to enchantment. Does that not invite their neo-Darwinian theorisation and bioscientific/biomedical manipulation and exploitation? Again, no. Such objectivising abstraction (without which such enterprises would be impossible) partakes wholly of 'enlightened' modernity, with its disenchanting effect; so it cannot coexist with the body and the material that is integral to enchantment. For that, a different kind of truth is required: the body as active agent and as lived, as well as living – of which Maurice Merleau-Ponty, Luce Irigaray and David Abram are the pre-eminent philosophers.[42] (Not, I would add, Foucault, for whom bodies merely passively bear and reproduce the power-relations inscribed upon them.) In this construal, subjectivity is never disembodied, while the material is, in Val Plumwood's words, 'already full of form, spirit, story, agency and glory'.[43] I would add that in such a world, the appropriate (one might say almost say 'default') mode is what it always has been: neither theism nor atheism, but animism.[44]

One implication of all this is that enchantment will always slip the nets prepared for it, and even as it sometimes occurs where and when it is invited (principally by artists of all kinds but also, in a qualified way, in religious ritual), it will continue to appear where and when it is not. For as Latour points out, *contra* Weber's worst fears, we have never been entirely or permanently modern. (To be that would amount, I take it, to psychosis.)

Nonetheless, we cannot rest in that blithe truth. I believe that upon pain of continuing in our present destructive and self-destructive course, we urgently need to rediscover and honour enchantment in the world and in our lives. Without that, all the scientific research, policy statements, committees and non-governmental organisation in the world will not suffice to establish a right relationship with the Earth and our fellow Earthlings. And that is my conclusion, except before closing I want to touch upon one example and enter a couple of important provisos.

Climate change: changing the solution

One example is climate change. Most basically, after decades of Intergovernmental Panel on Climate Change (IPCC) assessments, the Kyoto Protocol and various G8 conferences, emissions are still rising. As Mike Hulme comments, 'Perhaps this particular way of framing climate change (as a mega-problem awaiting, demanding, a mega-solution) has led us down the wrong road.' Quite, and the still more spectacular failures of a single, universal and enlightened carbon-market, or political treaty, or geoengineering intervention, or mass spiritual conversion await in the wings. Perhaps, as he adds, in place of a 'universalised and materialised climate change... we must now particularise and spiritualise it.'[45]

But let me sharpen the issue by pointing out that the 'enlightened' attitude I am criticising operates on the side of the angels too. For example, William Rees, who developed the concept of 'ecological footprint', despairs that 'intelligence and reason are not the primary determinants of human behavior'. Rather, 'brutish passion and instinct often overwhelm the godly gift of reason'. Rees realises that it is the 'economic growth paradigm' ('industrial capitalism') that is 'wrecking the ecosphere', but he attributes that to the 'biological drivers' of our 'lower' brain centres.[46]

From my point of view, this attribution is less plausible than the one I have put forward here, namely of said economic system as a perverse realisation of 'reason'. Beyond that, what are Rees's rationalist exhortations but the lineaments of anthropocentrism, the very structure of values and ideas of human exceptionalism and privilege that is implicated in every upwards ratchet of ecocide? And as such, in ecological terms, a spectacular failure? What will 'save' us, if anything, is not what apparently separates us from other animals but a conscious recognition and revaluing of what we share: the true commons, and common good, of our embodied and embedded life as Earthlings.

Another example (which, like the preceding, deserves more elucidation than I can give it here) is genetic engineering. Hark to the words of an influential scientist, writing in its early days:

> The old dreams of the cultural perfection of man were always sharply constrained by his inherited imperfections and limitations.... The horizons of the new eugenics are in principle boundless – for we should have the potential to create new genes and new qualities yet undreamed of.... For the first time in all time, a living creature understands its origin and can undertake to design its future.[47]

Now what about this programme, as laid out here, is not in the spirit of the Enlightenment – not, indeed, its rhetorical fulfilment?

I would like to add that whenever cultural justifications for horrors are produced (female genital mutilation, the mass slaughter of songbirds, bull-fighting, etc.), the contemporary progressive response is usually twofold: to insist that (1) the crimes can only be indentified and addressed thanks to universal Enlightenment values, and that (2) the solution is to overpower local cultural dynamics with the same. The first proposition, however, is nonsense, as one counter-example alone should suffice to show: there is an entire, venerable and profound tradition, entirely uninfluenced by the Enlightenment – namely, Buddhism – the foundational value of which is compassion and the relief of suffering.[48] As for the second assumption, if I am right, then any such attempt is doomed either to fail or to 'succeed' imperialistically; whereas the more hopeful strategy (although also without guarantees of success, of course) is to locate, articulate and strengthen countervailing *local* cultural values.

Two provisos

The first proviso is that I am decidedly not arguing for a new universal metaphysics according to which the world is 'really' enchanted. The point noted about the disenchanting effect of monist and universalist object-ivism, whether 'spiritual' or 'material', applies here too. Enchantment is a personal experience (whether individual or collective) – which is to say, embodied and embedded – or it is nothing. In an account that bears much repeating, the anthropologist Irving Hallowell, interviewing an old Ojibwe man by the Berens River in northern Manitoba, Canada, asked him, '"Are *all* the stones we see about us here alive?"' (Formally, in the Ojibwe language, they are.) 'He reflected a long while and then replied, "No! But some are."'[49] That is, could be, in lived life. For the assertion that *everything* is *necessarily* alive, merely inverting our currently dominant view to the contrary, is no improvement; the authoritarianism of a universalist mode remains untouched.

The second proviso is that enchantment is by its nature not only unpredictable but is also intermittent, temporary and/or incomplete. As they say of love, it lasts forever while you are in it: but only then. Boundaries and limits may be crossed, but they are not eliminated. (This is why enchantment has always been regarded by religions with universalist ambitions as inferior or counterfeit.) Put another way, as indigenous myth and folklore recognises, you cannot live in *faërie* (and remain human) forever. Indeed, the prerequisite for a healthy (non-grasping/ non-addictive) relationship to enchantment is, paradoxically, a strong ego and the ability to handle disenchantment.

Jan Zwicky's analysis is helpful here. She identifies the 'lyric' as experiences of wordless clarity and beauty, which deepen rather than transcend specificity, contingency and vulnerability. She contrasts this with the 'technological', a world of use-values, resources and manipulation, and then a mediating third mode: the 'domestic', which 'accepts the essential tension between lyric desire and the capacity for technology.... To become domestic is to accept that one cannot live in wordlessness. This is compatible with wanting to'. (It also permits kinds of use that differ from exploitation.)[50]

The upshot of these two provisos is that for purposes of practical philosophy (and I follow Wittgenstein in holding all philosophy to be ultimately practical), what we need to develop is not exactly a philosophy of enchantment but one which can accommodate it without analysing, reducing or explaining it away in terms of something else. More: a philosophy which encourages a *modus vivendi* that recognises and values enchantment.

Renewing humanism

Following Toulmin's lead again, I would like to suggest one promising candidate for such philosophy (metaphysics, ethics, politics). It is not the only one – others include ecofeminism, civic republicanism and communitarianism, as well as philosophical Daoism[51] – but it can certainly hold up its head in such company. I am thinking of Michel de Montaigne's Renaissance humanism: sceptical, in the true – that is, classical – sense of the word, not the arrogant dogmatics of scientism; tolerant, not patronisingly but from a genuine recognition of the existence and importance of others; and, above all, humane, without its object being limited to humans. Equally, it is difficult to read Montaigne's essays without noticing their acceptance of the reality of natural or we might say 'ecological' limits, but also of the reality of the sacred. Any such philosophy must indeed be both ecological and post-secular.[52] Yet he is often critical of institutionalised religion, as we too must be, notably its effects on indigenous peoples; and he writes respectfully, sometimes ruefully, of embodied life, sex and (within fifteenth-century European limits that no longer constrain us) of women. In all these respects, Montaigne offers a model that contrasts tellingly, point by point, with that of both the secular modernists and their reactionary anti-modernist opponents, the 'One World' of both resistance-is-futile economic globalisers and global religious imperialists. It also contrasts with the arrogant, masculinist and anthropocentric enterprise, with its techno-cornucopian faith in unending progress, which has assumed the mantle of humanism today.[53]

Of course there is a place for reason in such a philosophy. But intellectualism has badly misled us about reason. As against the modernist fantasy of ultimate control, to which rationalism has all too easily lent itself, Toulmin points to the unavoidability of plurality, ambiguity and uncertainty – qualities of life that classical and Renaissance scepticism emphasised – and the humility they entail. He advocates recognising and revaluing four kinds of practical knowledge that the modernist counter-revolution, beginning three and a half centuries ago, has displaced and suppressed: 'the oral, the particular, the local, and the timely.'[54] Bodiment, embeddedness and enchantment are of the same family. And they have enormous positive potential, to which I hope this paper will contribute its mite.

Notes

1   For further discussion and detail, see my *Ecological Ethics: An Introduction*, 2nd edn (Cambridge: Polit y Press, 2011).

2   In order to ensure easy access to this text, I have used one of the translations available on the internet: www.columbia.edu/acis/ets/CCREAD/etscc/kant.html (accessed 10.9.2015). No page numbers are supplied.

3   M. Foucault, 'What is Enlightenment?' pp. 32–50 in P. Rabinow (ed.), *The Foucault Reader* (Harmondsworth: Penguin Books, 1984).

4   Matthew Taylor, 'Twenty-First Century Enlightenment', London: RSA, 2010. Available online at: www.thersa.org/about-us/rsa-pamphlets/21st-century-enlightenment (accessed 28.9.14).

5   Kant, as per ref. 2.

6   Kant, as per ref. 2.

7   Kant, as per ref. 2.

8   Kant, as per ref. 2.

9   Kant, as per ref. 2.

10   S. Toulmin, *Cosmopolis: the Hidden Agenda of Modernity* (Chicago, IL: University of Chicago Press, 1990).

11   Foucault, 'Enlightenment': 38, 39.

12   Foucault, 'Enlightenment': 37.

13   B. Williams, *Ethics and the Limits of Philosophy* (London: Fontana Press, 1993), p. 18.

14   Foucault, 'Enlightenment': 38.

15   B. Latour, *We Have Never Been Modern* (Hemel Hempstead: Harvester, 1993), p. 36.

16 Foucault, 'Enlightenment': 35.
17 Foucault, 'Enlightenment': 42.
18 Foucault, 'Enlightenment': 45.
19 Foucault, 'Enlightenment': 43, 44.
20 Foucault, 'Enlightenment': 42.
21 Foucault, 'Enlightenment': 43.
22 Foucault, 'Enlightenment': 48.
23 Taylor, see ref. 2 above. See also M. Bunting, 'Comment', *The Guardian* (14.6.2010).
24 See Gray's excellent *Enlightenment's Wake*, 2nd edn (Abingdon: Routledge, 2007).
25 Taylor, as per ref. 4.
26 H. H. Gerth and C. Wright Mills (eds), *From Max Weber: Essays in Sociology* (London: Routledge, 1991): e.g., 220, 298–9, 331.
27 Weber: 155.
28 H. H. Gerth and C. Wright Mills (eds), *From Max Weber: Essays in Sociology* (London: Routledge, 1991), pp. 155, 282.
29 See A. Szakolczai, M. Weber and M. Foucault: Parallel Life-Works (London: Routledge: 1998).
30 V. Plumwood, *Environmental Culture: the Ecological Crisis of Reason* (London: Routledge, 2002).
31 See the work of S. Sullivan, e.g. 'Ecosystem Service Commodities – A New Imperial Ecology?', *New Formations* (2010) 69: 111–28.
32 Respectively: S. Toulmin, *Return to Reason* (Cambridge MA: Harvard University Press, 2001), p. 178; B. Herrnstein Smith, *Contingencies of Value. Alternative Perspectives for Critical Theory* (Cambridge MA: Harvard University Press, 1988), p. 149; and, with added irony, the Khmer Rouge.
33 E.g., Foucault, 'Enlightenment': 43.
34 Quoted in *University Publishing* 13 (1984), p. 15.
35 R. Acampora, *Corporal Compassion: Animal Ethics and Philosophy of Body* (Pittsburgh, PA: University of Pittsburgh Press, 2006).
36 Z. Bauman, *Intimations of Postmodernity* (London: Routledge, 1992), p. x.
37 M. Dickinson, personal communication; thanks also to Anthony Thorley for 'unbiddable'. On the wild, see G. Snyder, *The Practice of the Wild* (Berkeley, CA: Counterpoint, 1990).
38 See S. Kane, *Wisdom of the Mythtellers* (Peterborough: Broadview Press, 1998).
39 Weber: 282.
40 E. Viveiros de Castro, 'Exchanging Perspectives. The Transformation of Objects into Subjects in Amerindian Cosmologies', *Common Knowledge* 10:3 (2004) 463–84, p. 470. The second remark was made in the course of four lectures Viveiros de Castro gave on the same subject in Cambridge in 1998.
41 See the classic discussion in J. R. R. Tolkien, 'On Fairy-Stories', pp. 9–73 in *Tree and Leaf* (London: Unwin Hyman, 1988) [1964]).
42 See my 'Revaluing Body and Earth', pp. 41–54 in E. Brady and P. Phemister (eds), *Embodied Values and the Environment* (London: Springer, 2012).
43 Plumwood, *Environmental Culture*, p. 226.
44 See G. Harvey, *Animism: Respecting the Living* World (New York: Columbia University Press, 2006).
45 M. Hulme, *Why We Disagree About Climate Change. Understanding Controversy,*

Inaction and Opportunity (Cambridge: Cambridge University Press, 2009), pp. 330, 333.

46   W. E. Rees, 'Are Humans Unsustainable by Nature?' Trudeau Lecture (28.1.09).

47   R. Sinsheimer, 'The Prospect of Designed Genetic Change', *Engineering and Science* (April 1969) 8–13, p. 11.

48   See D. E. Cooper and S. P. James, *Buddhism, Virtue and Environment* (Aldershot: Ashgate, 2010).

49   I. Hallowell, 'Ojibwe Ontology, Behavior, and World View', pp. 19–52 in S. Diamond (ed.), *Culture in History: Essays in Honour of Paul Radin* (New York: Columbia University Press, 1960), p. 24.

50   J. Zwicky, *Lyric Philosophy* (Toronto: University of Toronto Press, 2004), pp. 258, 534.

51   For a good recent discussion, see the relevant chapters by V. Plumwood, A. Dobson and R. Eckersley respectively in A. Dobson and R. Eckersley (eds), *Political Theory and the Ecological Challenge* (Cambridge: Cambridge University Press, 2006), and *Daodejing: "Making This Life Significant": A Philosophical Translation*, transl. and ed. R. T. Ames and D. L. Hall (New York: Ballantine Books, 2003). See also Gray, *Enlightenment's Wake*, pp. 274–6.

52   See my 'Post-Secular Nature: Principles and Politics', *Worldviews: Environment, Culture, Religion* 11:3 (2007) 284–304.

53   See D. Ehrenfeld, *The Arrogance of Humanism* (Oxford: Oxford University Press, 1981), C. Lasch, *The True and Only Heaven: Progress and its Critics* (New York: W. W. Norton, 1991), and T. Brennan, *Globalization and its Terrors. Daily Life in the West* (Routledge 2003). On Montaigne, see D. L. Schaefer, *The Political Philosophy of Montaigne* (Ithaca, NY: Cornell University Press, 1990).

54   Toulmin, *Cosmopolis*, p. 30.

Chapter 6

(Re)embodying which body?
Philosophical, cross-cultural and personal reflections on corporeality

Sian Sullivan

Corporeal... *a.* bodily, physical, material; **corporeality**... *n.*
Embody... *v.t.* make (idea etc.) actual or discernible; (of thing) be an
expression of; include, comprise; **embodiment** *n.*[1]

Modernity's 'great divides'

In the seventeenth century, as is well known, a certain French philosopher conceptually divided the world into two disparate substances:
matter (*res extensa*, or 'extended substance') and mind (*res cogitans*, or
'thinking substance'). Matter, as Rene Descartes described it, was
spatially extended, determinate, and mechanical; mind, on the
contrary, had no spatial presence whatsoever – it was pure thought,
free of all physical constraint and limitation. While animals, plants, and
indeed nature as a whole were composed exclusively of mechanical
matter, and while God consisted entirely of mind, humans alone –
according to Descartes – were a mixture of the two substances. The
human body, like other animal bodies, was a completely mechanical
configuration; the immaterial mind somehow interacted with this
physical body from a location within the human brain.[2]

A general theme and even *raison d'être* of this collection of essays is that
multiple contemporary crises in the *ecosocius* can be traced to the 'great
divides' so defining of the modern era: between 'West' and 'Other', humans
and 'nonhumans', and culture and nature.[3] Ushered in by the long
Cartesian 'moment' of the 1600s, itself rooted in Renaissance interpretations of classical Greek philosophy[4] and particularly the ancient break
from *mythos* to *logos*,[5] this period in Enlightenment Europe saw an intensified
decoupling of ideas regarding mind and culture from those concerning
body and nature. As transcendent, disembodied minds were elevated over
proliferating abstractions of mechanised bodies – automata[6] – from cellular
to cosmic scales, western psyches arguably became increasingly detached

from the human and earthly bodies with which they are embedded and entangled.[7] Human nature has thus been rendered increasingly deaf to a stilled and desacralised more-than-human nature that is its mirror,[8] entrenching an Aristotelian position that 'Man' alone is a political animal, with nature-beyond-the-human rendered as politically mute.[9]

Combined with the privileging of calculative rationality[10] and positivist modes of verification[11] as rules for generating valid knowledge, this idealistic transcendence of mind from matter characterises the modern disembodying impetus of current concern. Signalling both a disconnect between mind and matter, and a hierarchisation that sets abstract thought above and over a projection of a machine-like body, this impetus deepens a corresponding set of binaries, that are also hierarchised.[12] As summarised in Table 1, we find ourselves living in the shadow of two thousand years of hierarchical value-ordering in western thought.[13] Through this, a reality has been constructed and normalised whereby only humans, and often only some humans, possess intelligence and mind; and where at the other end of the hierarchy, plants, for example, are viewed merely as 'vegetables'[14] – dispossessed of the capacities of movement, perception, communication, and self-directed *telos*, and thus usefully backgrounded as existing only for the instrumental ends of humans. In this hierarchical ontology only the intelligence characteristic of (particular) human entities can confer moral considerability, since only this intelligence is understood as possessing scope for communication, purpose, and subjectivity.

Table 6.1 Plato's and Aristotle's value hierarchies of the faculties of soul

Plato		Aristotle	
Spirited	Enabling activity and volition	Intellective	Rational soul possessing mind/reason 'human excellence'
Rational	Enabling intelligence and self-control. Associated with reason/ mind/opinion and located in men who are thus able to rule	Locomotive	Mobility found in humans and animals but *not* plants
		Desiderative	Able to desire, i.e. to have appetite, passion, wish – found in humans and animals
Appetitive	Associated with pleasure/ pain/desire as well as passivity.	Perceptive	Able to sense pleasure and pain – found in humans and animals
	Located in the ruled – slaves, women, children and slaves. Plants as fixed, rooted, passive	Nutritive	Mechanical ability to feed and reproduce. Plants possess only this 'soul', i.e. otherwise rendered as passive.

Source: Based on M. Hall, 2011;13 after V. Plumwood, 2006.

The ontological denial of these latter faculties in other kinds of embodied being, including, historically, in the bodies of the non-western Other,[15] permits the doing of harm without *recognition* that harm has been done. Although often it is more complex than this, in that the denial of capacities for communication, purpose, and subjectivity in 'non-human others' perhaps manifests more as *disavowal*: as the simultaneous acknowledgement of harms caused, accompanied by a strategy – an apparent solution – to seemingly mitigate this harm.[16] An early example of this, and of the pathology that such 'solutions' can embody, comes from the post-Cartesian vivisectionists. Whilst operating in a Cartesian mode, i.e. construing animals as soulless automata, these scientists would also cut the vocal cords of their experimental subjects so that they would not be able to hear the ensuing cries of pain.[17] Through this apparent 'solution' their embodied acknowledgement of the communicative and experiential capacities of animals was denied, so as to literally make the animals subject to their experiments into mute objects. Strategies of disavowal – of the simultaneous acknowledgement of, and turning away from, harms caused – abound today through the sale and purchase of various forms of tradable 'offsets' for 'solving' problems of environmental harm. Purchase of environmental 'credits' generated in one place are thus considered to 'solve' damage effected somewhere else, although arguably such 'solutions' also entrench a disconnection (or splitting-off) from the continuation of damage-producing behaviours that such offsets require.[18]

The effects of modernity's incomplete movement towards ontological divides between mind and matter, culture and nature, West and Other, masculine and feminine, then, are widely understood to be a denigration of bodily, sensuous, and ecological grounds for knowing and feeling.[19] As observed elsewhere in this volume (see, for example, chapters by Harris, Holden, and Young), this is a denigration that frequently also targeted women and non-Europeans as categories of humans considered to be closer to 'the body' and to 'nature'. In parallel, the production of a nature-beyond-the-human[20] that is distant, stilled, and 'outside', has created this nature as usefully amenable to objectification, instrumentalisation, and myriad associated violations.[21] Nonetheless, this semblance of control – this thinking that 'things stand mute and inert' until 'modern man' chooses to speak of them[22] – is frequently accompanied by dismay, as the materiality and unruly agency of natures-beyond-the-human burst through in the environmental and social fall-outs of industrial processes, so as to require corrective and frequently costly responses.

Given the *disembodying* hypostasis of contemporary pathologies in socionatural configurations, a reconnecting and compositionist[23] 'politics of nature' frequently advocates practices and strategies of *(re)embodiment*, echoed in calls for academic scholarship to (re)insert corporeality as the always present ground of experience.[24] Such corrective movements are understood in part to encourage choices towards expressions of corporeality that (re)acknowledge the material basis of being: 'the inescapable consequence of our physicality' known through the feelings of our bodies;[25] and 'the language of the earth, and ... of our bodies'.[26] Through this acknowledgement, (re)connections between human and earthly bodies are to be engendered so as to constitute correctives to the excesses and violences constituting the abstracted life and labour so necessary for industrial capitalism.[27] As such, this (re)embodying impetus tends to urge some sort of (re)turn to, and/or embrace of, 'the body' that was left behind by the disembodied Cartesian mind of the Enlightenment. It is suggestive of a re-entwining of 'bodymind' – a remembering that body and psyche 'are not separate entities but mysteriously a totality'[28] – combined with more attentive ecological attunement of human 'bodyminds' with the embodied entities comprising more-than-human natures.

At the same time, however, understandings, experiences, and performances of 'the body' are also historically and culturally situated, caught within specific regimes of truth that make possible particular embodiments and embodied experiences.[29] Acknowledgement of the particular mind–body split associated with the Cartesian moment in itself affirms that specific, culturally inflected, historically situated socialisation processes, associated with language learning, come into play to variously shape human encounters with, and understandings of, both corporeality and 'nature-beyond-the-human'.[30] Indeed, reversion to the mechanistic body-as-automata as conceived by Descartes would surely be contradictory to the hopeful, healing thrust of current calls for re-embodiment. The query in my title –'(re)embodying which body?' – reflects the productive complexity and ambiguity conferred through locating 'the body', corporeality and embodiment as always both caught within, as well as performing, the particularities of historical and cultural contexts.

In what follows, I respond to this query with some particular philosophical, cross-cultural, and personal gestures, the latter, denoted in indented italics,[31] based on fragments of my own memories of particular experiences of corporeality. My intention is to highlight the creative force of contradictions contained in the categories 'the body', 'corporeality', and '(re)embodiment' that, as noted by de Lucia (this volume), acknowledges that 'the body' is 'epistemically plural' as well as politically situated. A diversity of corporealities – understood, experienced, and performed bodies – is thus always at play in worldly and world-making participations. This includes any normative call for (re)embodiments resistant to the disembodying impetus of capitalist symbolic and material orders.

From docile to diffuse bodies: poststructuralism and corporeality

1987 – Discipline. For almost every day of ten years of professional dance training I have 'done morning class'. I see myself in the mirror, extending into an arabesque – all lines and taut coherence – punctuated by closures into neat fifth positions: a parody of mimesis through which I perform my mirrored body in this daily ritual of ballet school confinement. Sometimes, all this disciplined bodily participation permits the exhilarating sense of freedom that may emerge from repetitively practised technique. But often, control and composure mask the effort, sweat, and sometimes physical pain that is the ground of my experience.

discipline produces subjected and practised bodies, 'docile' bodies. Discipline increases the forces of the body (in economic terms of utility) and diminishes these same forces (in political terms of obedience).[32]

You will be organised, you will be an organism, you will articulate your body – otherwise you're just depraved. You will be signifier and signified, interpreter and interpreted – otherwise you're just a deviant. You will be a subject, nailed down as one, a subject of the enunciation recoiled into a subject of the statement – otherwise you're just a tramp.[33]

A gift of poststructuralist thought has been the foregrounding of the ways that historical contexts, comprising particular configurations of human and other-than-human natures, confer possibility and constraint to the known and the experienced. The gendered, sexualised, normal, and mad body; the lived body whose immanent and heterogenous life, via consolidated power relations, can be abstracted and calculated as exchangeable general labour;[34] the body that can be punished directly by sovereign powers, or self-controlled through the biopower sustaining élite political economy constellations: these shifting corporealities are always indelibly entangled with regimes of truth that potentiate particular embodied knowledges and experiences.

Michel Foucault, in particular, has encouraged a methodical approach, involving archaeologies of discursive knowledge production and genealogies of institutional *dispositifs*, to enable identification of associated objects of knowledge, technical and calculative renderings, and the interplay of discursive and institutional configurations that create and accompany 'regimes of truth'.[35] For Foucault, 'the body', that seemingly most material ground of being, is fully caught, inscribed, and repetitively enacted[36] within discursive webs, whose diagnosis may encourage awareness and contestation. While discourse structures and performs multiplicitous and networked details that can sediment into institutional *dispositifs* and relatively stable assemblages,[37] change becomes possible through reflection and critical understanding of historically empowered scripts shaping the acts that perform social reality.[38] As Foucault writes in his essay 'The subject and power',[39] we need to check both 'the type of reality with which we are dealing' *and* 'the historical conditions which motivate our conceptualization' of this type of reality, if we wish to both understand and adjust the power-effects that are thereby amplified.

This, then, offers a perspective on the possibility for change to emerge from embeddedness in particular regimes of truth, without recourse to the abstract-universal certainty (or Truth) assumed by Enlightenment science's desire for a decontextualised, objective view-from-everywhere-and-nowhere.[40] Key amongst Foucault's insights is that bourgeois, capitalist political economy requires certain sorts of labouring bodies that thus have (had) to be made through the 'conduct of conduct' («conduire des conduites») comprising an associated and consolidating liberal governmentality.[41] In *Discipline and Punish*, for example, Foucault emphasised that new regimes of governance were structured and bolstered by new social sciences of the modern era, which recursively and productively reinforced new disciplining techniques of management and administration.[42] The rise of the bourgeois class and the Age of Reason in Europe thus was accompanied by a novel emphasis on partitioning, classifying, codifying, and calculating 'the body'.[43] It is this calculative emphasis that constructs, subjects, manages, and accumulates corporeality as both a fungible source of homogenised abstract labour, and as utility-maximising 'body-machines', whilst simultaneously permitting the rationalisation and administration of seemingly docile, manageable bodies as populations.[44]

Arguably, the dissociative effects of these processes of subjectification through which bodies 'learn to labour'[45] have frequently been exacerbated by experiences of trauma. The workhouses, mines, slaveries, and genocidal colonial wars accompanying the modern will to industrial capitalism were (and are) run-through with individualised and (un)predictable violences:[46] shocks that sever a subject from the already known, creating impossibilities that cannot be psychically integrated.[47] These circumstances are relevant to a call for 're-embodiment' because the docile body organised so as to labour, may also be a traumatised body: a body that experiences itself as 'foreign' in its spontaneous impulses to surrender and flow; a subject who has learned it is 'vital to suppress feeling and emotional expression'[48] – to neutralise affect and memory so as to withdraw from the dangers and experiential vulnerabilities posed by engaged embodied existence.[49] Critiquing the societal normalisations informing much psychiatric practice, in *The Politics of Experience* R. D. Laing thus famously wrote that '[t]his state of affairs represents an almost unbelievable devastation of our experience':

[o]ur capacity to think, except in the service of what we are danger-ously deluded in supposing is our self-interest... is pitifully limited: *our capacity even to see, hear, touch, taste and smell is so shrouded in veils of mystification that an intensive discipline of unlearning is necessary for anyone before one can begin to experience the world afresh, with innocence, truth and love...*[50]

Movements towards a freer embodiment, attuned to the flourishing of diverse human and more-than-human bodies, thus intrinsically require a political process of unlearning the docile, labouring, traumatised body of such utility for the accumulations driving capitalist political economy.[51]

Under the *neoliberalism* of recent decades, corporealities have become increasingly entangled in what Thomas Lemke, following Foucault, identifies as the 'consistent expansion of the economic form to apply to the social sphere, thus eliding any difference between the economy and the social'.[52] This movement has been accompanied by the transposition of 'economic analytical schemata and criteria for economic decision-making onto spheres which are not, or certainly not exclusively, economic areas'.[53] Foucault's insights here are key in part because they extend to the recent emergence of a neoliberal 'environmentality'[54] that promotes and makes necessary a neo-liberal conduct of conduct in the sphere of socio-ecological relations.[55] Proliferating new conceptions of nature-beyond-human thus conceptually entrain these natures with 'market-based solutions' based on ownership of newly configured commodity units such as 'ecosystem services', 'biodiversity offsets', and 'carbon credits'.[56] As I have observed elsewhere,[57] nature-beyond-the-human is thereby made docile through a conceptual transformation that seeks to catch it 'in a [new] system of subjection', whereby its productive characteristics can be further 'calculated, organized, technically thought' and 'invested with power relations'.[58] Like the human body, and the body-politic of populations, nature refigured as a provider of 'ecosystem services' and a 'bank of natural capital' is 'entering a machinery of power that explores it, breaks it down and rearranges it to productively bend and release its immanent forces towards economic utility'.[59] The distributed neoliberal microphysics of *biopower* that both makes and rules corporeality and subjectivity thus 'not only regulates human interactions' and 'seeks directly to rule over human *nature*'.[60] It also extends to nature-beyond-the-human so as to become a technology of power over multiplicitous bodies that 'makes live and lets die':[61] a politics of life itself.[62]

2000 – Ecstasy. A different kind of repetition. Opening into rhythms and illicit spaces. The feeling of freedom in the 'temporary autonomous zones'[63] of raves and squat parties. Gatherings of souls similarly alienated by the boundaries, regulations, violences and inequities of formal society. In this urgent dance my heart opens, my soul soars, my consciousness seems simultaneously to permeate through and fly from all the cells of my body. I feel connected – improvisationally open to the moving bodies gathered here. Senses bathed in stimulation. Synaesthetically fused, perceptually expanded. Fully alive, and at home in this wild, weird, industrial world.

Every single molecule that Charis is taking into her lungs has been sucked in and out of the lungs of countless thousands of other people, many times. Come to that, every single molecule in her body has been part of somebody else's body, of the bodies of many others, going back and back, and then past human beings, all the way to dinosaurs, all the way to the first planktons. Not to mention vegetation. We are all a part of everybody else, she muses. We are all a part of everything.[64]

When you will have made him a body without organs, then you will have delivered him from all his automatic reactions and restored him to his true freedom.[65]

If Foucault affirms diagnosis of the structuring epistemic and institutional grids shaping corporeal understanding and experience as essential for appropriate ethical and other responses,[66] then poststructuralist philosophers Deleuze and Guattari, in their resolutely non-totalising affective ontology of rhizomes, nomadism, and becoming, offer conceptual strategies for exploding these structuring grids. Inspired by avant-garde writer, performer and theatre director Antonin Artaud to whom the term 'body without organs' (BwO) is attributed (see above), as well as by the burgeoning anti-psychiatry movement of 1960s and 1970s, their 'plateau' 'November 28, 1947: How do you make yourself a Body without Organs?' approaches 'the body' as *diffuse* – as 'fluid-multiple-open'[67] – rather than as (only) organised and particulate. For Deleuze and Guattari, experiential 'lines of flight' permit escape from the hyper-regulated and docile mind–body spaces naturalised under the striations of modernity's State Science.[68] Corporeality here makes possible transgression of the organised bodies constituting abstract labour and utility-maximising consumption, so as to shatter – through experiences that activate transgressive affective intensities – the neat illusions maintaining the status quo. Deleuze and Guattari frequently fetishise practices of pain, negation, and absence as the routes towards these transgressive, experiential potencies. At the same time, the world they encounter – of multiplicity, ephemeral becoming, and metamorphosis – encourages a certain optimism towards the possibility for transformation, invention, and reconstitution through improvisational participations in the 'plane of proliferation, peopling and contagion' they understand as the dynamically immanent basis of being/becoming.[69] As they ask, '[w]hy such a dreary parade of sucked-dry, catatonicized, vitrified, sewn-up bodies, when the BwO is also full of gaiety, ecstasy, and dance?'[70]

Arguably, then, the BwO of poststructuralist philosophy, and the turn to affect more broadly in conceptions of corporeality, embodiment, and subjectivity,[71] introduces a vital infusion of mind and intelligence throughout the materiality of the body. This is an intelligent and vibrant materiality that simultaneously extends to and connects with the diverse other and distinctive bodies, not to mention vibrant matters,[72] constituting nature-beyond-the-human. Bodies are thus invoked not 'as stable things or entities, but rather as processes which extend into and are immersed in worlds', to form aspects of dynamic 'assemblages of human and non-human processes'.[73] The BwO, and the consciously and unconsciously experienced lines of flight generating variously playful and improvisational intensities with other resonant bodies,[74] is thereby conceived and known as intrinsically and consciously ecological. Guattari, after Gregory Bateson,[75] thus iterates 'ecosophy' – 'an ethico-political articulation…' between the three ecological registers of 'the environment, social relations and human subjectivity' – as epistemologically necessary so as to re-embed relationships between interior (subjective) and exterior (social and environmental) potencies.[76]

Invocations of (re)embodiment(s) that contest the disembodying impetus of modernity's 'great divides', clearly require attention to the particular corporealities constituting the realm of possibility for this productive, but problematic, impetus. Invoking such attention perhaps shifts the call away from a simple encouragement for (re)embodiment, and towards identification of the specific corporealities expressed and proliferating in particular regimes of truth, so as to assert their identifiable power-effects in socio-ecological registers. This positioning recognises the mind-full corporeality – the intelligent physicality and materiality – of bodies that is always present as the ground of being; whilst acknowledging that context as well as choice consolidates particular embodied expressions and experiences of this corporeality. It is an affirmation that *difference makes a difference* in understanding, performing, and manifesting corporeality. This theme is taken further in the next section to draw attention to the significance of diverse cultural inflections regarding corporeality, in combination with the infinitely different 'demands' effected by the sheer diversity of embodied natures-beyond-the-human.

Difference makes a difference: cultural–corporeal inflections and affective animisms

2007 – Pain and healing. I'm lying under a lone shepherd tree, sun bright on the white bark that gives Boscia albitrunca *its distinctive name. I've promised myself 48 hours alone in the red desert west of the Dâures massif in Namibia where I've worked on and off for more than twenty years. My intention is to fully face the pain spread throughout my body. A burning that made its presence known following an exhilarating summer several years previously spent in street protests for social and environmental justice, with the exposure to tear gas toxins and other violences that this entailed.*[77] *I dive deep into the pain and into what feels like a jagged 'arrowhead' wedged deep between two vertebrae, and wonder where and how I will find a healer able to see and remove this festering source of discomfort. Unexpectedly, some months later I am with friends in a clearing in the upper Amazon, drinking a potent brew of the complex plant medicine* ayahuasca. *Don Luca, the 'doctor', reads me as he weaves a gentle* icaros *or spirit song into my consciousness, the rhythmic, cleansing movement of the leaves of the* chacapa *fan drawing my experience into the mysterious breathing of the forest. He says, 'you have pain in your body. It comes from a desert spirit that you did not know how to protect yourself from, and it will feel jagged like broken glass, just here', pointing to the exact place that I feel to be the source. And then, 'I can remove it – for $20'. Following this extraction that is also an exchange, I wander to the edge of the clearing where the vegetal complexity of the forest pulses with a life force of vibrant colours, iridescent in the dark. I join this dance, arms light like rainbow butterfly wings. For the first time in years I am free of physical pain.*

She is the one who knows the organs, if they move from one side to the other side. And there is one thing here in the middle – if you touch here he kicks. If that *!arab* is moving you just push a little bit push push, so that it can come in the middle. [...]

After we've given birth, after three months the child must be cut. Now those small tattoos, it's kind of preventing the sicknesses to come – like child sicknesses or paralysis. If the child is not cut they get those sicknesses. That's why we are cutting them with the tattoos and we put in the *abu||horo* – the ostrich egg shell – and the kudu skin into the tattoos. After that we take that black material and in that black material we put 5 cents, plus there are two things we are putting in there – one is a *≠âis* – it's a thought type of plant – it acts to protect against bad thoughts of the people. *≠Âis* is a *soxa* plant. These are very special plants. They will not work if you don't give something to them in exchange. And we also put in – they said it's an underground rhino – the ant lion – *!gudubes*. [...]

Abraham Ganuseb and Andreas !Kharuxab told me that when they are dancing the *arus* something is talking into their head and told them what to use [for healing]. We ask them how they know all these plants that they are using for stomach pain, coughing, everything, and they told us that some of the plants, the old people (*kai khoen*) used them, but some of the things they get from the spirit which is talking with them when they dance...[78]

If poststructuralist philosophy complexifies corporeality – drawing attention to how 'the body' and embodiment are always caught within regimes of truth influencing what can be known, experienced and contested – then cultural difference generates further disturbances in approaching bodily and earthly ground. Re-embodiment thus may not be as simple as resuscitating the intelligent materiality of human and other-than-human bodies, so as to counter the detachment effected by Cartesian abstract thought. This is because, and as illustrated in the testimonies above, cultural diversity seems also to generate corporeal as well as affective diversity. Or at least, once again, to shape the truths by which 'the body', corporeality and nature-beyond-the-human can and ought to be known, practised and performed.[79] This means that whilst it seems appropriate to acknowledge that 'there cannot help but be some overlap between my direct, visceral experience and the felt experience of other persons',[80] these very visceral and felt experiences are themselves filtered through ontological assumptions shaped by cultural (as well as personal) specificity. Despite the apparent universals of biology, different peoples of the world may express and know different conceptions of corporeality and embodiment.

To add further complexity, the diversely different embodied forms of multiple selves constituting nature-beyond-the-human create depth[81] and demand variation in embodied and cultural engagement with these selves. Cultures relying closely for sustenance on neighbourhood natures know this very well: the seeds of particular food plants require just the right amounts of rain and sun, and the agency of diverse predators – from slugs to elephant – are the focus of much attention and varied manipulations. The material agencies of a series of non-human fallouts of industrial processes may demand even closer anthropogenic attention: from radioactive waste committing economical and ecological attention for unimaginable numbers of future generations,[82] to decaying and indigestible plastic detritus causing death by starvation through clogging the alimentary tracts of numerous seabirds.[83] Approaches to natures-beyond-the-human may range from utilisation, appreciation and empathy, to ambivalence and detachment – even horror, disgust, and fear,[84] and all of these are also culturally inflected.

As Gregory Bateson[85] and more recently Eduardo Kohn[86] have emphasised, *difference makes a difference*. This is a truism that nonetheless requires acknowledgement, given the disruptions of cultural difference accompanying the universalising tendencies of modernity,[87] and its extension in modern totalisations of nature-beyond-the-human as Nature, 'biodiversity', and now 'ecosystem services' and 'natural capital'. There is a mirror here in the familiar significant cultural divide between the conceptually pacified and scientifically knowable Nature of the Enlightenment, and the lively, cacophony of selves-beyond-the-human known by modernity's necessarily animist Other. As the Nobel Laureate and molecular biologist Jacques

Monod wrote in the 1970s, science necessarily 'subverts every one of the mythical ontogenies upon which the animist tradition... has based morality' so as to establish 'the objectivity principle' as the value that defines 'objective knowledge itself'.[88]

An oft-invoked corrective to such reasoning thus is of a (re)invigorated amodern experience of the alive sentience of other-than-human-natures as animate and relational subjects, rather than inanimate and atomised objects. As such, 'animism' is a descriptor that enfolds Edward Tylor's 'mistaken primitives', positioned prior to the attainment of Enlightenment rationality in his theory of religion,[89] with postmodern 'eco-pagans' of the industrial west, for whom animism is a contemporary eco-ethical 'concern with knowing how to behave appropriately towards persons, not all of whom are human'.[90] It is both 'a knowledge construct of the West',[91] and a universalising term acknowledging a 'primacy of relationality',[92] as well as a set of affirmative, affective, and mimetic practices that 'resist objectification' by privileging an expansionary and reciprocal intersubjectivity.[93] Such relational connectivities, elaborated further by Harris and Holden (this volume), beautifully bind 'the embodied self' with multiplicitous other selves. In a systemic reversal of the rugged individualism of Enlightenment and liberal thought, it is perhaps only in full acknowledgement of this binding that the human 'self' can be experienced as fully 'free'.[94]

> *2014 – Sand, stars and fire. 'You are demanding to be cooked', says Khall'an, pulling me up once more, and drawing me back towards the women's voices. The complex polyphonic rhythms and harmonies, the driving sharp repetitions of the women's handclaps, the feeling of my shaking body pressed close against the shaking bodies of Khall'an and Tsama, coalesce so that I am falling to the sand beneath my feet, all senses simultaneously open to the stars of the Kalahari sky above. Astounded by this finely-tuned yet improvisationally open series of affective intensities, I understand why anthropologist Megan Biesele describes this bodily-mindfull-socio-ecological attunement as 'one of humanity's great intellectual achievements'.*[95]

The rest is silence. Corporeality – embodiment – death

> *2015 – Mortality. I write these lines as the 2015 London marathon is being run. At the first London marathon in 1981, my father ran the 26 miles in three hours. Today his days are marked by effort to maintain upright dignity as his neurological system descends into the misfiring and paralysis of Parkinson's Disease. In witnessing this inevitable corporeal succumbing to the 'suck' of entropy*[96] *I feel overwhelmed by the bittersweet poignancy of our variously*

disciplined efforts for mind to shape matter. Our worldly efforts to make manifest the diverse becomings of our dreams, knowing always the inevitability of their loss to us.

Vast in its analytic and inventive power, modern humanity is crippled by a fear of its own animality, and of the animate earth that sustains us. [...] [the] difficult mystery of our own carnal mortality... our vulnerability, our utter dependence upon a world that can eat us.[97]

In this chapter I have suggested that instead of emerging as a source of clarity regarding how to act and be in the world, both corporeality and embodiment are ambiguous, shifting, and dynamic affairs – sites of the continual recognition and negotiation of values. Nowhere does this ambiguity arise so strongly as in the inevitable mortality that is handmaiden to the embodied 'thingness' of corporeality. The intractable materiality of our individual bodies cannot help but be a reminder of an unavoidable connection: between the gift of awake embodiment that permits the flame of consciousness to flicker, if only briefly in this world; and the particulate earthly body, coalesced from cosmic sources and on a one-way street to return to these sources. We might direct mind over matter throughout life, but as we labour into death the entangled relationships between human and earthly bodies clearly are unavoidable.

Death is also a culturally inflected transition in which the ways that we have lived these entangled socio-ecological relationships become illuminated: drawing into focus our impacts and our offerings; our kindnesses, connections, and felt commonalities with each other and with natures-beyond-the-human. If 'we' feel a sense of kinship with the spirited materialities of earth to which we let go the individual moments of our lives, perhaps this labouring into death also becomes a return – a final homecoming – of our organic, nourishing corporeality to the embrace of both our human ancestors and the persistent (in)organic matter of life. In this acknowledgement, then, it is not only in embodied life that we might learn to live well with each other and with more-than-human-natures. It is also in approaching mortality as the inevitable consequence of this life, so as to (re)embody death, as well as life, well.

Acknowledgements

I am grateful for the support of the UK's Arts and Humanities Research Council (AH/K005871/2), and to Mike Hannis for close reading and comments on an earlier draft. The views expressed herein are mine alone.

Notes

1   Definitions from H.G. Fowler and F.G. Fowler, *The Pocket Oxford Dictionary of Current English*, 7th ed., R.E. Allen (ed.), Oxford: Clarendon Press, 1984, pp. 162, 239.

2   D. Abram, *Becoming Animal: An Earthly Cosmology.* New York: Pantheon Books, 2010, p. 103.

3   B. Latour, *We Have Never Been Modern.* Cambridge, MA: Harvard University Press, 1993, p. 97.

4   C. Merchant, *The Death of Nature: Women, Ecology and the Scientific Revolution.* New York: Harper and Row, 1989(1980).

5   S. Žižek, *Living in the End Times.* London: Verso, 2011, p. 279; P. Feyerabend, *Conquest of Abundance: A Tale of Abstraction over the Abundance of Being.* Chicago, IL: University of Chicago Press, 1999.

6   Thus in *Discourse 5* of Rene Descartes' *Discourse on Method* (London: Penguin Books, 1968(1637), pp. 75–76) he writes of animals that '…they do not have a mind, and… it is nature which acts in them according to the disposition of their organs, as one sees that a clock, which is made up of only wheels and springs, can count the hours and measure time more exactly than we can with all our art'. Other authors argue against the thesis that Descartes considered animals to be incapable of feeling, whilst affirming his insistence on animals as automata, possessing neither thought or self-consciousness (see P. Harrison, 'Descartes on animals', *The Philosophical Quarterly*, 1992, vol. 42(169), pp. 219–227).

7   See especially D. Abram, *The Spell of the Sensuous: Perception and Language in a More-Than-Human World.* London: Vintage Books, 1996; D. Abram, op. cit., 2010.

8   M. Weber, *The Protestant Ethic and the Spirit of Capitalism.* Abingdon: Taylor and Francis, 2001(1905); B. Latour, *Politics of Nature: How to Bring the Sciences into Democracy.* Cambridge, MA: Harvard University Press, 2004; P. Curry, 'Nature post-nature', *New Formations*, 2008, vol. 26 (Spring), pp. 51–64; D. Abram, op. cit., 2010.

9   A. Dobson, 'Democracy and nature: speaking and listening', *Political Studies*, 2010, vol. 58(4), pp. 752–768. Thus Indonesian indigenous narratives tell of '*the falling silent of the world* under the burden of "primitive accumulation," of capitalist exploitation, and of colonial administration', A. Franke, 'Animism: notes on an exhibition', *E-flux*, 2012, vol. 36, Available online at: www.e-flux.com/journal/animism-notes-on-an-exhibition/ (accessed on 2 January 2013), pp. 12–13, emphasis added.

10   M. Weber, 2001(1905), op. cit.

11   B. Latour, *An Inquiry Into Modes of Existence: An Anthropology of the Moderns.* Cambridge, MA: Harvard University Press, 2013.

12   D. Abram, op. cit., 2010, pp. 46–47.

13   Discussed in detail by philosophers M. Hall (*Plants as Persons: A Philosophical Botany*, New York: Suny Press, 2011) and M. Marder (*Plant-Thinking: A Philosophy of Vegetal Life*, New York, Columbia University Press, 2013). Also see D. Abram, op. cit., 2010, pp. 159–181.

14   M. Marder, 2013, op. cit.

15   See, for example, M. Taussig, *Shamanism, Colonialism and the Wild Man: A Study in Terror and Healing*. Chicago, IL: University of Chicago Press, 1987.

16   On 'disavowal' see S. Freud, 'Splitting of the ego in the process of defence', pp. 3–6 in Bokanowski, T. and Lewkovitz, S. (eds) *On Freud's 'Splitting of the Ego in the Process of Defence'*. London: Karnac Books, 2009(1938); R. Fletcher, 'How I learned to stop worrying and love the market: virtualism, disavowal, and public secrecy in neoliberal environmental conservation', *Environment and Planning D: Society and Space*, 2013, vol. 31(5), pp. 796–812; S. Sullivan, 'At the Edinburgh Forums on Natural Capital and Natural Commons: From disavowal to plutonomy, via "natural capital"', 2013a, Available online at: http://tinyurl. com/oj8l8e6 (accessed on 2 May 2015).

17   A. Hornborg, 'Animism, fetishism, and objectivism as strategies for knowing (or not knowing) the world', *Ethnos*, 2006, vol. 71(1), p. 24, after N. Evernden, *The Human Alien: Humankind and Environment*. Toronto: University of Toronto Press, 1985, pp. 16–17.

18   R. Fletcher, 2013, op. cit.; S. Sullivan, 2013a, op. cit., S. Sullivan, 'After the green rush? Biodiversity offsets, uranium power and the 'calculus of casualties' in greening growth'. *Human Geography*, 2013b, vol. 6(1), pp. 80–101.

19   D. Abram, op. cit., 1996, 2010; V. Plumwood, *Feminism and the Mastery of Nature*. London: Routledge, 2006.

20   A brief note on terms used here to denote 'nature(s)'. 'More-than-human nature' is advocated by phenomenologist David Abram as a way of overcoming the way that 'nonhuman nature' defines nature-beyond-the-human in a negative sense, i.e. as nature-that-is-not-human (D. Abram, op. cit., 1996). Abram's intention is to acknowledge that the *human world* is always a subset of the latter, but never the other way around. The human world thus is always *'embedded within, sustained by* and thoroughly *permeated by,* the more-than-human world', while the more-than-human world, although including the human world, and frequently 'profoundly informed by the human world', 'always *exceeds* the human world' (Abram, D. pers. comm.). I also use the terms 'other-than-human' nature(s) or 'nature-beyond-the-human' (after Kohn, E. *How Forests Think: Towards and Anthropology Beyond the Human*. Los Angeles, CA: University of California Press, 2013), when referring to organisms, entities and contexts other than the modern common sense understanding of the biological species *Homo sapiens*. At the same time, the situation may be even more complex. This is because in many 'animist' and amodern cultural contexts embodiments other than the modern biological species category of *Homo sapiens* may be perceived ontologically as representing different bodily perspectives – different natures – that nonetheless are embraced by a broader, inclusive category of human persons (see Viveiros de Castro, E. 2004 Exchanging perspectives: the transformation of objects into subjects in Amerindian ontologies. *Common Knowledge* 10(3): 463–484; also S. Sullivan and C. Low, 'Shades of the rainbow serpent? A KhoeSān animal between myth and landscape in southern Africa – ethnographic contextualisations of rock art representations', *The Arts*, 2014, vol. 3(2), pp. 215–244). Invoking 'nonhuman' or 'more-than-human' nature in these cultural contexts might thus discount the perceptual and ontological reality guiding understanding and practice in such contexts, in which a greater degree of underlying ontological,

communicative and *cultural* continuity is acknowledged between different embodiments 'in nature' than might be the case in the species thinking informing modern natural science.

21   As described, for example, in T. Ingold, 'Rethinking the animate, re-animating thought', *Ethnos*, 2006, vol. 71(1), pp. 9–20; D. Abram, op. cit., 2010; P. Descola, *Beyond Nature and Culture*. Chicago, IL: University of Chicago Press, 2013; M. Marder op. cit., 2013.
22   D. Abram, op. cit., 2010, p. 40.
23   After B. Latour, An attempt at a 'compositionist manifesto', *New Literary History*, 2010, vol. 41, pp. 471–490.
24   T. Csordas, *Embodiment and Experience: The Existential Ground of Culture and Self*. Cambridge: Cambridge University Press, 1994.
25   D. Abram, op. cit., 2010, pp. 19, 46; after M. Merleau-Ponty, *Phenomenology of Perception*, C. Smith (trans.). London: Routledge and Kegan Paul, 1962(1945).
26   D. Jensen, *A Language Older Than Words*. New York: Context Books, 2000, p. 311.
27   B. Banerjee, 'Necrocapitalism', *Organization Studies*, 2008, vol. 29(12), pp. 1541–1563.
28   M. Starks Whitehouse, 'The tao of the body', pp. 41–50 *in* P. Pallaro (ed.) *Authentic Movement: Essays by Mary Starks Whitehouse, Janet Adler and Joan Chodorow*. London: Jessica Kingsley Publishers, 1999(1958), pp. 41–42.
29   M. Foucault, *Discipline and Punish: The Birth of the Prison*, trans A. Sheridan. London: Penguin, 1991(1975); J. Butler, 'Performative acts and gender constitution: an essay in phenomenology and feminist theory', *Theatre Journal*, vol. 40(4), 1988, pp. 519–531. A. Mol, *The Body Multiple: Ontology in Medical Practice*. London: Duke University Press, 2002.
30   D. Abram, 2010, op. cit., p. 39; E. Kohn, 2013, op. cit.
31   Also see O. Jones, 'An ecology of emotion, memory, self and landscape', pp. 205–218, in J. Davidson, L. Bondi and M. Smith (eds) *Emotional Geographies*. Oxford: Ashgate, 2005.
32   M. Foucault, 1991(1975), op. cit., p. 138. Also discussion in A. Dunlap, 'Permanent war: grids, boomerangs and counter-insurgency', *Anarchist Studies*, 2014, vol. 22(2), pp. 55–79.
33   G. Deleuze and F. Guattari, *A Thousand Plateaus: Capitalism and Schizophrenia*, B. Massumi (trans), London: The Athlone Press, 1987(1980).
34   M. Foucault, 1991(1977), op. cit.; D. Chakrabarty, *Provincialising Europe: Postcolonial Thought and Historical Difference*. Princeton: Princeton University Press, 2000, pp. 91–92 and references therein.
35   M. Foucault, *The Order of Things: Archaeology of the Human Sciences*. London: Routledge, 2001(1966); M. Foucault, *The Archaeology of Knowledge*. London: Routledge, 2002(1969).
36   cf. J. Butler, 1988, op. cit.
37   cf. B. Latour, *Reassembling the Social: An Introduction to Actor-Network Theory*. Oxford, Oxford University Press, 2005.
38   M. Foucault, *The Will to Knowledge: The History of Sexuality*, vol. 1, trans. R. Hurley. London: Penguin Books, 1998(1976).
39   M. Foucault, 'The subject and power', *Critical Enquiry* 8(4): 777–795.
40   cf. S. Žižek, 2011, op. cit., p. 282, discussing arguments made by D. Chakrabarty, 2000, op. cit.

41   M. Foucault, *Dits et écrits IV*. Paris: Gallimard, 1994, p. 237, see https://foucault blog.wordpress.com/2007/05/15/key-term-conduct-of-conduct/, last accessed 25 April 2015.

42   M. Foucault, 1991(1975), op. cit.

43   Ibid., pp. 137–138; also S. Federici, *Caliban and the Witch: Women, the Body and Primitive Accumulation in Medieval Europe*. New York: Autonomedia, 2004, chapter 4; S. Sullivan, 'Banking nature: the spectacular financialisation of environmental conservation', *Antipode*, 2013c, vol. 45(1), pp. 198–217.

44   M. Foucault, *Security, Territory, Population*. New York: Picador, 2007.

45   After P. Willis, *Learning to Labor: How Working Class Kids Get Working Class Jobs*. New York: Columbia University Press, 1981(1977).

46   cf. M. Taussig, 1987, op. cit.; B. Banerjee, 2008, op. cit.; C. Cavanagh and T.A. Benjaminsen, 'Virtual nature, violent accumulation: the "spectacular failure" of carbon offsetting at a Ugandan National Park', *Geoforum*, 2014, vol. 56, pp. 55–65; A. Dunlap and J.R. Fairhead, 'The militarisation and marketisation of nature: an alternative lens to "climate-conflict"', *Geopolitics*, 2014, vol. 19(4), pp. 937–961.

47   S. Žižek, 2011, op. cit., p. 292; C. Malabou, 'Post-trauma: towards a new definition?', pp. 226–238 *in* T. Cohen (ed.) *Telemorphosis: Theory in the Era of Climate Change*, vol. 1. Ann Arbor, MI: Open Humanities Press, University of Michigan, 2012.

48   N. Totton, 'Foreign bodies: recovering the history of body psychotherapy', pp. 7-26 *in* T. Taunton (ed.) *Body Psychotherapy*. Hove: Brunner-Routledge, 2002, p. 17.

49   S. Žižek, 2011, op. cit., p. 293–295, after C. Malabou, *Les Nouveaux Blessés (The New Wounded)*. Paris: Bayard, 2007.

50   R.D. Laing, *The Politics of Experience and The Bird of Paradise*. London: Penguin Books, 1967, pp. 22–23, emphasis added.

51   I have discussed these themes in fuller detail in S. Sullivan, 'On dance and difference: bodies, movement and experience in KhoeSān trance-dancing – perspectives of a "raver"', pp. 234–241 *in* W.A. Haviland, Gordon and L. Vivanco, L. (eds) *Talking About People: Readings in Contemporary Cultural Anthropology*, 4th Edition. New York: McGraw-Hill, 2006(2001); S. Sullivan, 'We are heartbroken and furious!' Rethinking violence and the (anti-)globalisation movements (#2)', *CSGR Working Paper* no. 133/04, 2004. Available online at: www2.warwick.ac.uk/fac/soc/csgr/research/workingpapers/2004/wp13304.pdf (accessed on 2 May 2015); Sullivan, S. 'An *other* world is possible? On representation, rationalism and romanticism in Social Forums', *ephemera: theory and practice in organization*, 2005, vol. 5(2), pp. 370–392. See also R. Fletcher, 'Free play: transcendence as liberation', pp. 143–162 *in* R. Fletcher (ed.) *Beyond Resistance: The Future of Freedom*. New York: Nova Science Publishers Inc., 2007.

52   T. Lemke, 'The Birth of Bio-Politics' – Michel Foucault's Lecture at the Collège de France on Neo-Liberal Governmentality', *Economy and Society*, 2001, vol. 30(2), pp. 190–207, p. 197.

53   T. Lemke, ibid.; S. Sullivan, 'The natural capital myth; or will accounting save the world? Preliminary thoughts on nature, finance and values', *LCSV Working Paper* 3, 2014. Available online at: http://thestudyofvalue.org/wp-content/

uploads/2013/11/WP3-Sullivan-2014-Natural-Capital-Myth.pdf (accessed 2 May 2015).

54   A. Agrawal, *Environmentality: Technologies of government and the making of subjects.* Durham, NC: Duke University Press, 2005; R. Fletcher, 'Neoliberal environmentality: towards a poststructuralist political ecology of the conservation debate', *Conservation and Society*, 2010, vol. 8(3): 171–181.

55   S. Sullivan, 'The elephant in the room? Problematizing 'new' (neoliberal) biodiversity conservation', *Forum for Development Studies*, 2006, vol. 33(1), pp. 105–135; B. Büscher, S. Sullivan, K. Neves, D. Brockington and J. Igoe, 'Towards a synthesized critique of neoliberal conservation', *Capitalism, Nature, Socialism*, 2012, vol. 23(2), pp. 4–30.

56   Discussed further in S. Sullivan, 2013c, op. cit. S.Sullivan, 2014, op. cit. On the new property rules required of market-based mechanisms for allocating environmental goods and bads, see C.T. Reid, 'Between priceless and worthless: challenges in using market mechanisms for conserving biodiversity', *Transnational Environmental Law*, 2013, vol. 2, pp. 217–233.

57   Sullivan, 2013c, op. cit., p. 211.

58   M. Foucault, 1991 [1975], op. cit., pp. 24–26.

59   M. Foucault, Ibid., pp. 138, 170.

60   M. Hardt and A. Negri, *Empire.* Cambridge, MA: Harvard University Press, 2000, p. xv, emphasis added.

61   M. Foucault, 1998(1976), op. cit.; M. Foucault, *The Birth of Biopolitics: Lectures at the Collège de France 1978–1979*, trans G Burchell. Basingstoke: Palgrave Macmillan, 2008(1979).

62   N. Rose, 'The politics of life itself', *Theory, Culture and Society*, 2001, vol. 18(6), pp. 1–30.

63   H. Bey, *TAZ: The Temporary Autonomous Zone, Ontological Anarchy, Poetic Terrorism.* New York: Autonomedia, 2011(1985, 1991).

64   M. Atwood, *The Robber Bride.* London: Virago, 1994, p. 56.

65   A. Artaud, 'To have done with the judgment of God', *in Selected Writings.* Susan Sontag (ed.). Berkeley, CA: University of California Press, 1976, p. 571.

66   M. Foucault, *The Care of the Self. The History of Sexuality, vol. 3.* London: Penguin Books, 1990(1984).

67   S. Žižek, *Organs Without Bodies: On Deleuze and Consequences.* London: Routledge, 2004, p. 4.

68   G. Deleuze and F. Guattari, 1987(1980), op. cit.

69   G. Deleuze and F. Guattari, Ibid., pp. 258, 266–7, 397–398, 426; see discussion in T. Mueller and S. Sullivan, 'Making other worlds possible? Riots, movement and counter-globalization', *in* Davis, M.T. (ed.) *Crowd Actions in Britain and France from the Middle Ages to the Modern World.* Basingstoke: Palgrave Macmillan, in press.

70   G. Deleuze and F. Guattari, 1987(1980), p. 150.

71   See L. Blackman, *Immaterial Bodies: Affect, Embodiment, Mediation.* London: Sage, 2012.

72   Cf. J. Bennett, *Vibrant Matter: A Political Ecology of Things.* Durham, NC: Duke University Press, 2010.

73   L. Blackman, 2012, op. cit., p. 1. Also B. Latour, 2005, op. cit.

74   B. Massumi, *What Animals Teach Us About Politics.* Durham, NC: Duke University Press, 2014.

75 G. Bateson, *Steps to an Ecology of Mind*. Chicago, IL: University of Chicago Press, 2000(1972).

76 F. Guattari, *The Three Ecologies*, Ian Pindar and Paul Sutton (trans), London, Continuum, 2000(1989), pp. 19–20.

77 S. Sullivan, 2004, op. cit.

78 Extracts from an interview with Welhemina Suro Ganuses in north-west Namibia, 3 March 2014. On KhoeSãn medicine, see C. Low, *Khoisan Medicine in History and Practice*. Köln: Rüdiger Köppe Verlag, 2007.

79 For anthropological perspectives on the diverse cultural variations and intellectual traditions regarding corporeality and 'nature', see especially C. Lévi-Strauss, *The Savage Mind*. Chicago, IL: University of Chicago Press, 1966(1962); C. MacCormack and M. Strathern (eds) *Nature, Gender and Culture*. Cambridge: Cambridge University Press; and E. Kohn, 2013, op. cit.

80 D. Abram, 2010, op. cit., p. 143. Also S. Sullivan, 2006(2001), op. cit. and C. Low, 'The role of the body in Kalahari San healing dances', *Hunter Gather Research*, 2015, vol. 1(1), pp. 27–58.

81 cf. D. Abram, 2010, op. cit., pp. 81–101.

82 As powerfully conveyed in the film *Into Eternity*, which documents the building of a geological storage facility for radioactive waste in Finland (www.intoeter-nitythemovie.com/); S. Sullivan, 2013b, op. cit.

83 M. Jackson, 'Plastic islands and processual grounds: ethics, ontology, and the matter of decay', *Cultural Geographies*, 2013, vol. 20(2), pp. 205–224.

84 V. Flusser, *Vampyroteuthis infernalis*. New York: Atropos Press, 2011(1987); F. Ginn, 'Sticky lives: slugs, detachment and more-than-human ethics in the garden', *Transactions of the Institute of British Geographers*, 2014, vol. 39(4), pp. 532–544.

85 G. Bateson, 2000(1972), op. cit.

86 E. Kohn, 2013, op. cit.

87 D. Chakrabarty, 2000, op. cit.

88 J. Monod, Chance and Necessity, trans. A. Wainhouse. Glasgow: Collins, 1972, pp. 160–164, quoted in M. Midgley, *The Myths We Live By*. London: Routledge, 2011(2004), p. 4.

89 E. Tylor, *Primitive Culture*, 2 vols. London: John Murray, 1913(1871); also G.W. Gilmore, *Animism or Thought Currents of Primitive Peoples*, Whitefish, Montana: Kessinger Publishing, LLC, 2010(1919).

90 G. Harvey, *Animism: Respecting the Living World*. London: Hurst and Co., 2005: xi; also A. Plows, 'Earth First! Defending Mother Earth direct-style', pp. 152–135 *in* G. Mckay (ed.) *DIY Culture: Party and Protest in Nineties Britain*. London: Verso, 1998; A. Letcher, '"Gaia told me to do it": resistance and the idea of nature within contemporary British eco-paganism', *Ecotheology*, 2003, vol. 8(1), pp. 61–84; A.P. Harris, *The Wisdom of the Body: Embodied Knowing in Eco-Paganism*. Unpublished PhD Thesis, University of Winchester, 2008.

91 H. Garuba, On animism, modernity/colonialism, and the African order of knowledge: provisional reflections. *E-flux*, 2012, vol. 36. Available online at: www.e-flux.com/journal/on-animism-modernitycolonialism-and-the-african-order-of-knowledge-provisional-reflections (accessed on 2 May 2015), p. 7.

92 N. Bird-David, '"Animism" revisited: personhood, environment, and relational epistemology', *Current Anthropology*, 1999, vol. 40(Supplement), pp. S67-S91; T. Ingold, 2006, op. cit.

93 J. Lewis, '*Ekila*: blood, bodies, and egalitarian societies', *Journal of the Royal Anthropological Institute*, 2008a, vol. 14, pp. 297–315; J. Lewis, 'Maintaining abundance, not chasing scarcity: the real challenge for the 21st century', *Radical Anthropology*, 2008b, vol. 2, pp. 11–18; A. Franke, 2012, op. cit., pp. 4, 7; E. Kohn, 2013, op. cit.; Discussed further in S. Sullivan, 'Nature on the Move III: (re)countenancing an animate nature', *New Proposals: Journal of Marxism and Interdisciplinary Enquiry*, 2013d, vol. 6(1–2), pp. 50–71.

94 M. Hannis, 'The virtues of acknowledged ecological dependence: sustainability, autonomy and human flourishing', *Environmental Values*, 2015, vol. 24(2), pp. 145–164.

95 Email to author, 9 May 2014, emphasis added.

96 D. Abram, 2010, op. cit., p. 48.

97 D. Abram, Ibid., p. 69.

Chapter 7

The knowing body
Eco-paganism as an embodying practice

Adrian Harris

Introduction

Any attempt at theorising re-embodiment needs to consider two questions: First, how did the Western experience of the self become 'essentially disembodied'?[1] Second, what might guide Western humankind to re-embodiment? Although the first question has been discussed at length, it cannot be left unconsidered, for the genealogy of disembodied experience carries vital clues for strategies of re-embodiment. So in the first section of this chapter I shall consider this question. I propose that our sense of disembodiment is closely related to the emergence of modern urban civilisation. Although we can trace the process from early history, I shall focus on what I consider to be the most profound and damaging shift – from a medieval to a modern consciousness.

The second section of this chapter will consider strategies for re-embodiment. Curiously enough, these emerge from the very ancient and the very modern: animism, a spiritual belief probably held by the earliest humans, and embodied situated cognition (ESC), a field so new it only emerged at the close of the twentieth century. Animism and ESC illuminate each other, as well as re-embodiment, because ESC is fundamental to human consciousness and animism is our default relationship to the world.

The terms 'animism', 'spirit' and 'embodiment' are used in many different contexts and can carry quite different connotations. I will therefore be as specific as space allows about how I use these terms.

The term animism has an unpromising ancestry as it was used by the Victorian anthropologist Tylor to refer to 'lower' forms of religion that held a 'belief in Spiritual Beings'.[2] For Tylor, who was embedded in a colonialist materialist worldview, this was a 'primitive' and 'childlike' error.[3] The term has been re-appropriated by recent scholars as 'the new animism'.[4] From this perspective '[a]nimists are people who recognise that the world is full of persons, only some of whom are human, and that life is always lived in relationship with others'.[5]

Following this discussion with a consideration of what we mean by 'spirits' serves well, as I have begun to contrast a dualistic worldview of matter and spirit with a more relational one that emphasises process. A modernist Western worldview like Tylor's sets 'spirits' apart from 'matter' just as it separates mind and body. This chapter proposes a sideways step to a quite different stance where spirits and matter are revealed as no more separate than mind and body are. In general, when Eco-pagans speak about the spirit of a place, they do not envision a material location somehow inhabited by an immaterial spirit entity; spirit is immanent in the world.

My understanding of embodiment draws primarily on the phenomeno-logical tradition, especially as articulated in the work of Merleau-Ponty. Embodiment is not concerned with the body as such but with our phenom-enological experience of our bodily being-in-the-world. We are embodied beings, and what we think, how we feel and ultimately perhaps *who we are*, emerges from our fleshy embodied existence.

Before proceeding, I will set out my core assumptions. First, many aspects of human consciousness are malleable and will vary depending upon the social and cultural context of peoples lives. As cognitive science shows, 'the experiences of the body-in-the-world also shape the embodied mind'.[6] Second, individual experiences of a sense of embodiment or apparent disembodiment rest upon a long spectrum and neither are an absolute or fixed condition. My degree of awareness of my own embodi-ment can vary considerably over a single hour, much less a lifetime, but considering historical processes necessitates working with social norms that generalise across broad populations. Having said that, I take it as a given that Western civilisation tends to create a sense of a disembodied self, and as a result we feel alienated from our own bodies and from the world.

From participation consciousness to disembodiment

How did we acquire this illusionary sense of disembodiment? To answer that question fully would require a lengthy history of somatic experience, such as that offered by Berman.[7] For my purposes it will suffice to recount the most recent and profound episode in this long saga, the shift from medieval to modern consciousness. This account will model and highlight the process and illuminate the bigger picture.

Several anthropologists, philosophers and historians have discussed the notion of participatory perception[8] or participatory consciousness.[9] Abram, drawing upon Merleau-Ponty and his own fieldwork, concludes that 'perception always involves, at its most intimate level, the experience of an active interplay, or coupling, between the perceiving body and that which it perceives'.[10] This mode of awareness is largely eschewed in the Western world and has been replaced by what Berman calls 'nonparticipatory consciousness': 'that state of mind in which one knows phenomena precisely in the act of distancing oneself from them'.[11] Non-participatory consciousness enables scientific objectivity but it also encourages – perhaps even requires – an alienation from embodied awareness.

The gradual shift from an embodying, embedding consciousness to an alienated, disembodying and objective one arguably started with Plato, but it took centuries. We can usefully take the medieval period as a mid-point in this process, when participatory consciousness, though probably 'not so pure as that of pre-Homeric Greece',[12] was still common. Berman notes that in many medieval practices 'the route to true understanding was to be found in that absorption, in the loss of psychic distance'.[13]

The embodying participatory consciousness of the medieval period engendered a 'deeply animistic' understanding.[14] As Berman explains, '[p]erception and cognition emerged primarily from the body which is why, to borrow a term from the anthropologists, everything possessed *mana*, was alive'.[15] This resulted in a 'complex syncretism between animism and Christianity' that lasted for a century and a half.[16] But the objectivity demanded by modern science destroyed that balance as it created a new form of consciousness.

We can map this process through Bakhtin's analysis of the shift from the open or grotesque body to the Modern closed body. The notion of the open body emerges in Bakhtin's description of the carnival, which played a dominant role in people's lives in Medieval Europe. In large cities as much as three months a year were devoted to carnival activities. These were festive public occasions in which the populace filled the streets, drinking and masquerading. Bakhtin characterises carnival as a celebration of the body and openness to life.

A central aspect of carnival is its attitude to the body, which Bakhtin called the 'grotesque'. Grotesque bodies are not closed, but are open to the world. Emphasis is placed upon body parts that can reach out, such as the nose, the belly, phallus and breasts, and upon openings such as the mouth, genitals and anus. The intent is to reveal the body as part of the world – not separate from it. 'This is an unfinished and open body without clearly defined boundaries'.[17]

Carnival temporarily created a different habitus – an 'open body' that is more connected to other bodies and the environment around it. Thus:

> People could have an experience of the lived body that was more direct and unmediated – a sensuous involvement with the world where the boundaries between the inside and outside of the body, and the dividing line between the individual and the collective, were not as sharply drawn as they are today.[18]

With the Renaissance came what Bakhtin calls the 'Classical Canon', at the core of which was the rationalism 'created and expressed in Descartes' philosophy'.[19] The objectivist philosophy espoused by Descartes, Newton and others divided mind from body and simultaneously mechanised the world. An 'increasing preoccupation with psychic distance can be seen in most areas of human activity from the Renaissance on'.[20] Berman notes the discovery of perspective in art, the shift from alchemy to chemistry and from astrology to astronomy. Parallel changes occurred across society and culture as bodies began to acquire a private quality, closed off to the world. At the same time there emerged notions of good manners that required careful control of behaviour. Western civilisation thus created a 'closed body' that was both disembodying and alienated.

I have, of necessity, simplified somewhat: There are complex socio-historical relationships between the emergence of scientific rationalism, heightened self-awareness and the resultant loss of participatory consciousness. My principle point is that the 'shift from animism to mechanism' engendered 'not merely a new science, but a new personality to go with it...'[21] That personality is disembodying, a 'closed and rationalized Cartesian body' that is 'severed from its sensual connections with the world and its collective associations with other beings'.[22]

The closed modernist body has been created by a specific set of circumstances and is thus open to transformation. Furthermore, research into the processes of human cognition suggests that such a closed body is disembodying and therefore maladaptive. There are many dimensions to this maladaptive disembodiment, but I shall focus briefly upon its impact upon our environment. Bateson outlined the belief system that inevitably emerges from the closed modernist body:

> you will logically and naturally see yourself as outside and against the things around you. And as you arrogate all mind to yourself, you will see the world around you as mindless and therefore not entitled to moral or ethical consideration.[23]

He concludes that:

> If this is your estimate of your relation to nature *and you have an advanced technology*, your likelihood of survival will be that of a snowball in hell.[24]

Re-embodiment would be a major corrective advance in human development. It can be encouraged by a better appreciation of human cognition and animism, arguably our default mode of relating to the world.

Embodied situated cognition

As the twentieth century grew old, many philosophers worked to banish the disembodying spectre of Cartesian dualism. At the heart of this project lay a new theory of cognition as embodied, situated and intersubjective. What have emerged are models of ESC that disrupt established discourses that 'divide spirit and flesh, soul and body, subject and object... people and environment'.[25]

Merleau-Ponty's phenomenological insights clarified the ways in which our consciousness is incarnate in the world. He writes:

> As I contemplate the blue of the sky... I abandon myself to it and plunge into this mystery, it 'thinks itself within me', I am the sky itself as it is drawn together and unified, and as it begins to exist for itself; my consciousness is saturated with this limitless blue...[26]

Our awareness does not emerge from a disembodied, decontextualised mind located somewhere outside the physical. Nor is this mind disembodying, decontextualising of other embodied forms. Rather, the mind is part of an *active relationship* between embodied humans and the world. Merleau-Ponty concluded that the process by which we come to understand the world emerges from a unity between subjects and objects that is the direct result of our embodiment. 'The properties of the object and the intentions of the subject... are not only intermingled; they also constitute a new whole'.[27] This understanding disrupts the Cartesian world-view, because it transcends subject/object dualism: The 'I' that thinks is tangled with the object that is thought about.

Merleau-Ponty's insights have been influential upon later philosophers and cognitive science, a multidisciplinary field that draws upon biology, chemistry, psychology, information science, philosophy, anthropology and linguistics. Second-generation cognitive science marks a shift towards re-embodiment and 'begins with the realization that the body... grounds and shapes human cognition'.[28] Contrary to centuries of dualist theorising, reason not only relies upon the structure of the brain and the body, but also depends on our entire social and physical environment. Cognitive science confirms what Merleau-Ponty realised: cognition is embodied and blurs conventional understandings of 'self' and 'world'.

Philosopher Clark draws upon the insights of cognitive science to answer a question that is central to this chapter: 'Where Does the Mind Stop and the Rest of the World Begin?'[29] Clark follows Bateson in concluding that what we normally accept as 'mental processes' sometimes extend beyond the 'skin bag' into the local environment,[30] and this blurring of mind and world may challenge Western notions of self. Such conclusions are widespread: In his survey of the field, Peterson notes that for a 'significant number of researchers... to understand the mind/brain in isolation from biological and environmental contexts is to understand nothing'.[31]

Varela and colleagues build upon Merleau-Ponty's work to develop a model of cognition as 'embodied action', a process they call 'enactive'.[32] They agree that cognition is embodied and factor in the wider 'biological, psychological, and cultural context'.[33] By emphasising action they highlight that cognition is an aspect of the sensory body and that 'knower and known, mind and world, stand in relation to each other through mutual specification or dependent coorigination'.[34] They conclude that 'organism and environment enfold into each other and unfold from one another in the fundamental circularity that is life itself'.[35]

Such ideas may initially seem strange to us because we rarely have any phenomenological experience of the enfolding of organism and environment. There is, however, a very common embodied tacit knowledge that partly draws upon this enfolding. Gendlin, who is both a philosopher and a psychologist, describes a 'bodily sensed knowledge',[36] which he calls the 'felt sense'. A felt sense is basically a feeling in the body that has a meaning for us – those fuzzy feelings that we don't usually pay much attention to. Perhaps the most familiar example is the vague feeling that *something* is wrong, but we cannot say what it is. At such times we might say 'I just got out of the wrong side of bed this morning'. We often talk about having a 'gut feeling' about something or someone and when that is literally a bodily sensation, it's a felt sense. In such situations there is a knowing and a *not* knowing at the same time. What is unknown in each case is tacit and embodied, but we can bring it into explicit conscious awareness using Focusing, a simple technique Gendlin developed to facilitate working with the felt sense.[37]

In common with the thinkers considered above, Gendlin's conception of the 'body' extends beyond the skin: For Gendlin the body can be best understood as 'an ongoing interaction with its environment'[38] such that the bodily knowledge of the felt sense *is* the entire situation. Thus, via the felt sense we can access 'a vast amount of environmental information' and unexpected knowledge – accurate 'gut feelings' for example – can emerge. Gendlin supports the conclusion of the thinkers I have introduced: In the light of our current understanding, the subject/object distinction collapses: 'We will move beyond the subject/object distinction if we become able to speak from how we interact bodily in our situations.'[39]

The enactive process model

By combining enactivism with Gendlin's philosophy of the implicit, I have synthesised a model of ESC with more explanatory power than either has alone. This model is consistent with other theories discussed here, and in several cases elucidates them. Given that Gendlin's key exposition describes his theory as *A Process Model*[40] I call this the *enactive process model*. The enactive process model claims that our being-in-the-world is bound up with

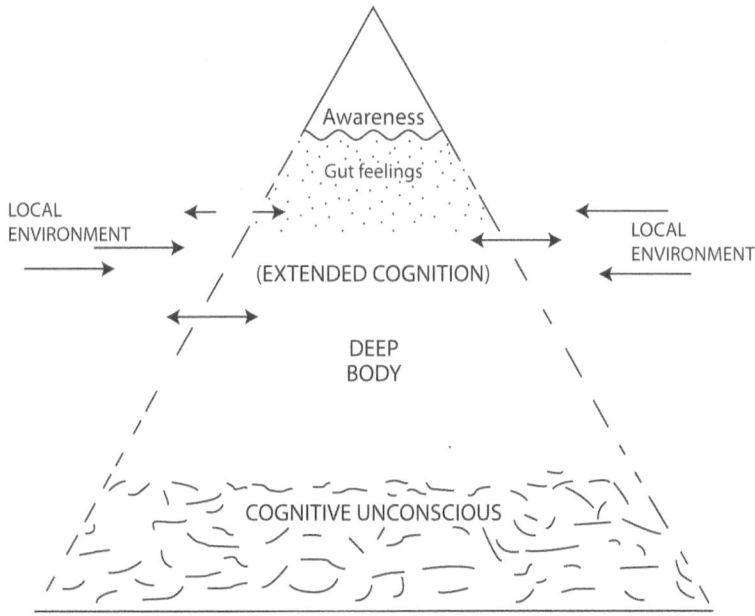

Figure 7.1 The cognitive iceberg

the immediate environment and that embodied cognition draws upon that space as a source of material to think with.

My 'cognitive iceberg' schematically represents the complex processes of ESC. It is inevitably an oversimplification and presents the local environment and physical body as more separate than the enactive process model suggests. I explain the enactive process model in detail elsewhere,[41] but, in summary, the whole 'iceberg' triangle represents the physical body, while the area at its base represents the 'cognitive unconscious'.[42] The physical body is engaged in a dynamic relationship with the local environment through perception and extended cognition. As 95 per cent of embodied thought occurs below our consciousness,[43] most of this processing never reaches everyday awareness, which is at the iceberg's tip.

At the top of the triangle – the tip of the proverbial iceberg – is everyday conscious awareness, which as we have heard, is a very small percentage of who we are. Consciousness is simply what we are aware of, the minimal aspects of a complex process, but because we identify our 'self' with consciousness we tend to discount what I call the *deep body* 'self' that actually governs much of our behaviour. This top level of awareness is quite narrowly focused and tends to heighten our impression of a subject/object distinction. The dotted area just below the apex designates 'gut feelings' or felt senses. Further down the triangle awareness widens out into the deep body, becoming less focused and blurring the distinction between self and other, shown in the graphic by the gaps appearing in the sides of the triangle. The cognitive unconscious is distinguished from the rest of the deep body because it is normally inaccessible to intentional influence or conscious awareness.

Applying the model

Certain circumstances and techniques allow us to become more aware of the blurred boundary between self and world. As the waves ride up the side of a real iceberg, what is above the water and what is below changes constantly. So it is with conscious awareness: At times we are unaware of the deeper processes of ESC – the sea around the iceberg is still. But at other times the sea is rough, and what lies beneath and above the waves shifts constantly. Our experiences make this apparent, as Leder vividly describes upon an occasion when he was walking in the woods, caught up with his own concerns:

> a paper that needs completion, a financial problem. My thoughts are running their own private race, unrelated to the landscape... The landscape neither penetrates into me, not I into it. We are two bodies.[44]

Leder's mind is working off-line,[45] and upon my model his awareness is focused at the tip of the cognitive iceberg. But the 'rhythm of walking' and the peace of the wood calm his mind and induce an 'existential shift', so that he begins to notice the beauty around him. Gradually

> [t]he boundaries between the inner and the outer thus become porous... I feel the sun and hear the song birds both within-me and without-me... They are part of a rich body-world chasm that eludes dualistic characterization.[46]

Leder's awareness has slipped down the cognitive iceberg, broadening into what Greenwood calls magical consciousness,[47] and this change in 'body-mind-habitus' produces 'an altered sense of self'.[48] A fundamental aspect of this change in habitus is the deepening sense of personal embodiment that results from shifting awareness down the cognitive iceberg. This shift blurs the distinction between self and other, enhancing Leder's sense of connection.

Berman suggests that in order to reach an embodied understanding of the premodern period 'we would have to abandon modern conscious-ness... and this means to abandon a certain type of egoic personality structure, allowing the mind to sink into the body, as it were'. He concludes that such a 'merger is premodern consciousness, or at least a good part of

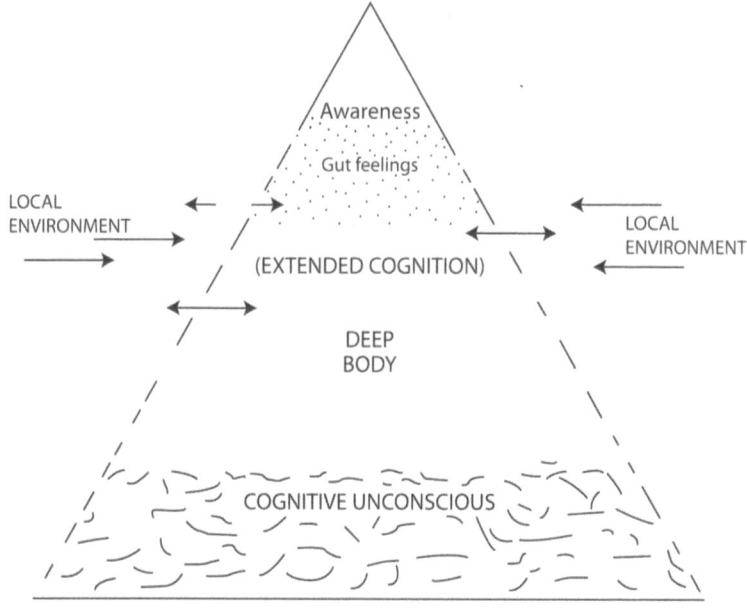

Figure 7.2 The premodern cognitive iceberg

it'.[49] The kind of premodern consciousness Berman describes can be represented with my 'Cognitive Iceberg' model as below. This differs from my model of modern consciousness in that the wavy line separating awareness from gut feelings is absent – the mind can sink into the body.

As awareness blurs into the extended cognition of the deep body, we would expect the kind of ease with participation consciousness I described earlier. It is probably not possible – or particularly desirable – to return to this form of premodern consciousness, but similar states of mind do exist in the modern world, notably amongst animists.

Animism – old and new

Abrams suggests that 'at the level of our spontaneous, sensorial engagement with the world around us, we are all animists', engaged in participation consciousness.[50] But how often are we aware of that sensorial engagement? An indigenous tribal lifestyle typically requires a high level of sensory awareness to survive. The Kaluli people of Papua New Guinea, for example, live in dense rainforest and rely upon a highly developed sense of sound to know the time of day, season of the year and their location in space.[51] It is no coincidence that indigenous tribal people are generally animists: Heightened sensory awareness encourages participation consciousness, which itself reveals the interconnectedness of the world. Abram asserts that such cultures are fully aware that we are 'corporeally embedded' in a 'living landscape'[52] and proposes that the body is 'a sort of open circuit that completes itself only in things, in others, in the encompassing earth'.[53] Ingold concurs that '"organism plus environment" should denote not a compound of two things, but one indivisible totality'.[54] Body and mind are thus simply different ways of describing the same process, 'namely the environmentally situated activity of the human organism-person'.[55] This mode of being-in-the-world engenders a knowledge that is grounded in 'practical application' and 'based in feeling, consisting in the skills, sensitivities and orientations that have developed through long experience of conducting one's life in a particular environment'.[56] Thus, as Bird-David explains, animism is grounded in 'a relational epistemology'.[57]

I do not want to idealise indigenous tribal peoples or imply that those living in a modern urban environment are incapable of participation consciousness: We are all capable of experiencing such consciousness. A fully embodied sensuality can break down the apparent division between body/mind/self and the 'other'. When we fully open to our direct sensory experience – the feel of the wind as it caresses the skin or the feel of the ground under our feet as we walk upon it – our awareness shifts. Abram notes that if one can maintain this awareness for some length of time, 'one will begin to experience a corresponding shift in the physical environment. Birds, trees, even rivers and stones begin to stand forth as living, communicative presences'.[58] This process is a development of the experience described by Leder, above, when '[t]he boundaries between the inner and the outer thus become porous'[59] Such blurring between 'inner' and 'outer' allows us to become aware that 'I' am 'that', and opens us to an animist ontology that entails a relational epistemology.

There are at least two fundamentally different ways of knowing: An embodying animist knowing that utilises participation consciousness and a disembodying Cartesian (modernist) scientific knowledge.[60] Bird-David compares the two: 'the object of modernist epistemology is a totalizing scheme of separated essences, approached ideally from a separated viewpoint', while the object of... animistic knowledge is understanding relatedness...'[61] While she concludes that '[b]oth ways are real and valid',[62] there are clearly costs in adopting a modernist epistemology.

Modernist epistemology is familiar to us and has been discussed endlessly, but embodied animist knowing has been virtually ignored. Part of the reason for that is the denial of the body that has infected Western philosophy since Plato, as encapsulated by Spelman's term, somatophobia.[63] There is a profound potential in embodying animist knowing for an emergent process of re-embodiment. We are nonetheless too soaked in our modernism to attempt an ersatz recreation of an indigenous oral culture. Western Eco-pagans, however, model a post-modern embodying animism.

Thinking with place: eco-pagans and other animists

My doctoral research focused on embodied knowing in eco-paganism. It demonstrated how the practice and belief of these modern Western animists draws upon ESC. In common with other animists, Eco-pagans feel a deep connection with the land, which is seen 'as sacred'.[64] One Eco-pagan (Jan) expressed it very explicitly: 'That's what paganism is all about – connection with everything'.[65]

Letcher suggests that 'an embodied sensitivity to nature' is essential if we are to 'come to know the *"genius loci"* – the spirit(s) of a place'.[66] This sense of connection to the *genus loci* functions through ESC; more specifically, it uses what I call processes of connection. All these processes tend to deepen our sense of connection with the natural environment and often alter our sense of self: they thus enable a communion with the *genius loci*, which informs Eco-paganism. I identified seven processes of connection during my fieldwork; the wilderness effect/threshold brook, meditation, the felt sense, trance, ritual, dance and entheogens.[67] All of the processes of connection can shift awareness down the cognitive iceberg, making the enactive process of co-creating reality described by Varela more apparent. This sometimes enables Eco-pagans to 'communicate' with the *genius loci* within which they are enactively enmeshed. There is only space to briefly consider each of these processes,[68] and I shall focus upon how they relate to embodiment in contemporary Eco-pagan animism.

The wilderness effect

Greenway argues that 'civilization is only four days deep'[69] and that even such a brief time away from twenty-first-century life is deeply transformative. The wilderness effect[70] was originally observed in the context of extended wilderness trips, but my fieldwork showed not only that the wilderness effect occurs at protest camps, but that it catalyses a spiritual experience that leads some protesters to describe themselves as 'pagans'.

There are many obvious similarities between a wilderness trek and life upon a typical protest camp. Most significant of these is the practical connection with the elemental forces of nature. As Letcher says: 'The very act of living out, however dependent on wider society for food and so on, puts one in touch with nature in a way that is real, not virtual'.[71] Newbury activist Jim Hindle describes how he

> became accustomed to the sound of the wind in the trees at all times. It wasn't a thing I necessarily listened to, but the silence that fell whenever I stepped inside a building was eerie and disquieting. ...It was like being connected to a great river, the source of all life... and years of separation between us and the Land were falling away like an old skin,[72]

Spending time in a more natural environment enhances sensory acuity[73] and this is apparent in my interview with Rob. In the city he has to engage sensory filters to 'block out information, to block out noise, to block out the chatter of things'. Upon returning to a more natural space he would find the silence overwhelming, but this passed:

> And it was only when you actually started to listen that you realised it wasn't quiet at all but the river was flowing, the wind was in the trees, the birds flying. All of these things were going on which we weren't hearing because we had these filters on. And I keep repeating it but it's an important point, because people do live their entire lives in an urban environment and they just don't get the connection, um they don't get that connection with nature.

For Greenway connection – or reconnection – is fundamental to the wilderness effect: 'When entering the wilderness psychologically as well as physically, participants most often speak of feelings of expansion or reconnection'.[74] The wilderness effect is closely related to the other processes of connection I introduced above, and all of them are grounded in embodied situated knowing: Shaw explains the sense of connection at the heart of the wilderness effect as 'an embodied visceral knowing that transcends the distinction between the inner and outer landscapes'.[75]

In an urban location the influence of the natural environment is diminished, and urban Eco-pagans need to enhance their sensory acuity to enable an animist connection to the *genius loci*. Such subtle embodied communion with place requires an intentional effort, which may involve several of the processes of connection. Urban Eco-pagans use meditation, entheogens, ritual and dance, trance and the felt sense to fine-tune their awareness of the natural environment and open up to the sensory richness available in intimate local relationships. This process of developing a deep connection to a specific location is most aptly referred to as 'listening to the threshold brook'.[76] As with more typical experiences of the wilderness effect, the threshold brook is deeply healing and can inspire a spiritual sensibility.

Meditation

Many Eco-pagans use some form of meditation. Eco-pagan Letcher recommends 'solitude, stillness and sensitivity' as a means of reconnecting with a natural environment.[77] This simple technique of spending time quietly 'sitting out' has a long history in Heathen[78] practice.[79] Greenway compares meditation with other processes of connection, claiming that 'both the psychedelic and meditation experiences... closely parallel'[80] the experience of the wilderness effect. He further claims that such awareness seems to have the 'capacity to open consciousness to Mind – that is, to the more natural flows of information from nature'.[81]

A sense of nature

Many Eco-pagans use felt senses in their spiritual practice. Barry provided an excellent example when he explained his animist 'pattern of consciousness' during my interview with him. I asked if there was a physical sense associated with this pattern:

> Barry: Definitely. Absolutely definitely... when you said the physical sense, and I focused there and I went there, I moved into that, I turned the volume up on that and I put my awareness into that, and what I got was a sense of embodiment which was much richer and more strange than your normal body awareness. Um, and so in a sense what happens is the hawthorn buds about to burst into May blossom became a physical sensation within my body... You know what I mean?
>
> Adrian: Kinda. Whereabouts in the body?
>
> Barry: OK. I'll do hawthorn buds. Oh, yeah, there is a practice I do that's related to this locating it in the body. Hawthorne buds are very much in my upper arms, and my chest. [Pause]. My shoulders. Like a kind of – You couldn't call it a buzz, I'm not talking about a buzz – I'm talking about a kinda, something, delicate.

Elsewhere Barry explains that a 'conversation with a tree is first and foremost a feeling in your body'.[82] This description of a felt sense enabling animist communication is common amongst Eco-pagans.

Situated embodied knowing in trance

Eco-pagan Shaman Gordon MacLellan writes of how the physical ecstasy of dance connects him to a 'world that thinks'.[83] Greenwood's model of magical consciousness[84] applies Bateson's model of an ecology of mind[85] to understand the way Gordon uses dance to communicate with the other-than-human world:

> Gordon's dance is about participating in such an interconnected system as an inspirited pattern – a web of wyrd – whereby the act of dancing enables spirits, energy and people to meet in a world that is alive.[86]

Bateson suggested that dance could serve as an 'interface between conscious and unconscious',[87] offering a means of understanding messages that the dancer is consciously unaware of. This does not involve a Freudian unconscious but one built upon an ecology of mind and is therefore consistent with the enactive process model. This process is apparent in Greenwood's description of an occasion when she watched Gordon dance with his spirit family:

> As the drumming increased, it was evident to me that there was a participatory communication between Gordon and the spirits in process, the other than human was coming into the human form. At times there seemed to be a non-verbal discussion going on as Gordon's body appeared to act out questions and answers in a swirling profusion of expressive movements.[88]

This is a good example of how trance might function within the enactive process model; by shifting his awareness to a deep embodied self, Gordon melts the boundary between subject and object, enabling communication with the 'other than human'.

Situated embodied knowing in ritual and dance

Dance can speak a language beyond words, and this reveals two inter-related aspects of ESC: First, dance allows a place or spirit to communicate to the dancer. Second, the dance can serve as a bridge between the dancers deep embodied knowing and their conscious awareness. These aspects work together when a place or a spirit communicates with the dancer through a deep embodied knowing that must be expressed in dance to become conscious. This process becomes apparent in the interview with Zoe when we read how her body has 'a different way of moving' that is 'like a dance' when she is connecting to a place. When Zoe's 'body moves then something is able to move' and this dance enables 'inspiration and the expression' to 'flow through'. The 'something' that moves is '[w]hat the place is trying to tell me', which can be 'an insight' or 'a sense of, my ancestors or the ancestors of the place being around me'. The *movement* of the dance is fundamental and without movement there is no contact: 'If [...] I'm frozen in some way physically, then I can't hear. I can't listen. Nothing will flow through me. So I have to move in some way with it, however simple it is, I have to move physicality in some way'.

Zoe experienced a similar sense when working ritually with a particular site:

> half way up the hill, there's an old hawthorn tree. So I would always stop there as I felt she was the guardian of the outer-ring. I've no idea if anyone else ever felt this, but I would have to stop there [...] ask her permission, and then wait for the answer. And the answer would come in a bodily sense. [...] It's like a sense of permission in my body.

This permission was sensed as a slight pressure on her back: 'It was like a propelling forward motion from behind. Like "Yes!" You know. Pushing me gently forward'. Once past this stage Zoe would come to 'the inner level' where she would often

> receive a 'no'. That would be like a frontal sense – like a closing down. I could feel it, yeah, it's definitely a front of body closed down. [Holds her open palms in front of her body]. Like someone's just drawn a curtain or shut a door in front of me.

Blurring boundaries

The processes of connection can dramatically reveal how 'organism and environment enfold into each other'.[89] Rob described how he felt one evening in the woods when a deep realisation of environmental destruction came to him:

> I felt like Gaia was really screaming out through me, saying please help me. Please help me, and like I started screaming myself and started saying these words. I felt so connected, so at one with the earth that this violence was being done towards me. Um, not me personally, any ego or anything like that, but me as in life, as in this whole unity which I'm connected with.

His identification with a sense of life itself, which is emphatically not his ego, is particularly striking and recalls Greenway's conclusion that some of what I call processes of connection 'facilitate the arousal of nonegoic awareness.'[90] Such experiences are not uncommon. Taylor found that 'no small number of activists report profound experiences of connection to the Earth and its lifeforms'[91] while Eco-pagan Jodie concluded that site life constructed 'a different form of consciousness whereby a person felt a part of nature'.[92]

Living upon a protest site – or wilderness trekking – will also have significant effects upon what Jackson describes as the 'body-mind-habitus'[93] thus contributing to changes in one's being-in-the-world. Jackson claims that changes in our habitus can free 'energies bound up in habitual deformations of posture or movement produce an altered sense of self',[94] which is exactly what the wilderness effect does. A fundamental aspect of this change in habitus is a deepening sense of personal embodiment, which I describe above in terms of shifting awareness down and through the cognitive iceberg.

Conclusion

My fieldwork touched the individual threads of many lives, and my writing has woven them into a tapestry with a distinct pattern: by various means we slip down the cognitive iceberg to become aware of 'a larger Mind of which the individual mind is only a subsystem'.[95–96] This plunge into the deep body awakens us from the dualistic dream that we are separate from the 'wisdom of the body'.[97] We can experience this psychological shift phenomenologically as a sense of spiritual connection that allows us to 'attune... to the natural world', and can feel

> [l]ike being in a great big dream, relevant messages are being spoken everywhere, telling me things I need to hear, and to which I need respond.[98]

Such messages were spoken in Gordon's dance and by the threshold brook, and we heard of their power to change lives. In as much as the immanent sacred is that which enables communion with the world and offers spiritual knowing, its source is the deep body which blurs into the natural environment. However deeply we drink from this source – a threshold brook perhaps – the depth of potential implicit knowing will never be drained and the experience of connection remains ineffable.

Eco-pagan practice can enable an awareness that embraces the reality of our embodiment and honours the processes by which self emerges from a relationship with the world. Although it may only appeal to a minority, understanding eco-paganism provides valuable theoretical insights for a cultural shift towards a richer sense of our embodiment.

Notes

1   I. Burkitt, *Bodies of Thought*, London: Sage Publications, 1999, p. 45.
2   E. Tylor, *Primitive Culture: Researches Into the Development of Mythology, Philosophy, Religion, Art, and Custom*, Volume 1. London: John Murray, 1871, p. 383.
3   E. Tylor, ibid., p. 431.
4   G. Harvey, *The Handbook of Contemporary Animism*, Durham: Acumen Publishing, 2013, p. 5.
5   G. Harvey. *Animism, Respecting the Living World*. London: Hurst & Co., 2005, p. xi.
6   T. Rohrer, 'The Body in Space: Embodiment, Experientialism and Linguistic Conceptualization', in J. Zlatev, T. Ziemke, R. Frank, R. Dirven (eds) *Body, Language and Mind*, vol. 2, Berlin: Mouton de Gruyter, 2007, p. 343.
7   M. Berman, *The Reenchantment of the World*, Ithaca, NY: Cornell University Press, 1981; M. Berman, *Coming to Our Senses*, London: Unwin, 1990.
8   D. Abram, *The Spell of the Sensuous*, New York: Pantheon, 1996; M. Merleau-Ponty, *Phenomenology of Perception*, London and New York: Routledge & Kegan Paul, 1962.
9   Berman, *Our Senses*, op. cit.

10 Abram, op. cit., p. 57.
11 Berman, *Reenchantment*, op. cit., p. 39.
12 Berman, *Our Senses*, op. cit., p. 122.
13 Ibid.
14 S. Harding, *Animate Earth: Science, Intuition and Gaia*, Totnes, Devon: Green Books, 2006, p. 25.
15 Berman, *Reenchantment*, op. cit., p. 133.
16 S Harding, op. cit., p. 25.
17 M. Bakhtin, *Rabelais and his World*, trans. by H. Iswolsky, Bloomington, IN: Indiana University Press, 1984, pp. 26–7.
18 Burkitt, op. cit., p. 46.
19 Ibid.
20 Berman, *Our Senses*, op. cit., p. 113.
21 Ibid.
22 Burkitt, op. cit., p. 49.
23 G. Bateson, *Steps to an Ecology of Mind: Collected Essays in Anthropology, Psychiatry, Evolution and Epistemology*, Chicago, IL: University of Chicago Press, 2000 [1974], p. 468.
24 Bateson, op. cit., p. 468.
25 G. Harvey, 'Animism: A Contemporary Perspective', in B. Taylor (editor-in-chief), *The Encyclopedia of Religion and Nature*, London and New York: Continuum, 2005, p. 83.
26 Merleau-Ponty, op. cit., p. 249.
27 Ibid., p. 13.
28 T. Rohrer, op. cit., pp. 21–2.
29 A. Clark, *Being There Putting Brain, Body, and World Together Again*, Cambridge, MA: MIT Press, 1997, p. 213.
30 Ibid., p. 214.
31 G. R. Peterson, *Minding God: Theology and the Cognitive Sciences*, Minneapolis: Augsburg Fortress, 2003, p. 43.
32 F. J. Varela, E. Thompson and E. Rosch, *The Embodied Mind: Cognitive Science and Human Experience*, Cambridge, MA: MIT Press, 1991, p. xx.
33 Ibid., p. 173.
34 Ibid., p. 150.
35 Ibid., p. 217.
36 E. Gendlin, *Focusing*, New York: Bantam, 1981, p. 25.
37 Ibid.
38 E. Gendlin, 'The Primacy of the Body, not the Primacy of Perception: How the Body Knows the Situation and Philosophy', *Man and World*, 1992, vol. 25(3–4), p. 349.
39 E. Gendlin, 'How Philosophy Cannot Appeal to Experience, and How it Can', in D. M. Levin (ed.) *Language beyond Postmodernism: Saying and Thinking in Gendlin's Philosophy*, Illinois: Northwestern University Press, 1997, p. 15.
40 E. Gendlin, *A Process Model*, New York: Focusing Institute, 1997. Available online at: www.focusing.org/process.html (accessed 30 August 2014).
41 A. Harris, *The Wisdom of the Body: Embodied Knowledge in Eco-Paganism*, PhD thesis, University of Southampton, 2008. Available online at: www.the greenfuse.org/phd/more/full_thesis.pdf (accessed 20 March 2013).

42   G. Lakoff and M. Johnson, *Philosophy in the Flesh: The Embodied Mind and its Challenge to Western Thought*, New York: Basic Books, 1999, p. 10.

43   Ibid., p. 13.

44   D. Leder, *The Absent Body*, Chicago, IL: University of Chicago Press, 1990, p. 165.

45   On-line cognition deals with live tasks that require fast moment-by-moment processing. We switch to slower, off-line cognition when we need to check upon something odd or plan future behaviour.

46   Leder, op. cit., pp. 165–6.

47   S. Greenwood, *The Nature of Magic: An Anthropology of Consciousness*, Oxford and New York: Berg, 2005.

48   M. Jackson, 'Knowledge of the Body', in H. Moore and T. Sanders, *Anthropology in Theory: Issues in Epistemology*, Malden, MA: Blackwell, 2006, p. 328.

49   Berman, M., *Our Senses*, op. cit., p. 123.

50   Abram, op. cit., p. 57.

51   S. Feld, 'Places Sensed, Senses Placed', in D. Howes, *Empire of the Senses: The Sensual Culture Reader*, Oxford and New York: Berg, 2005.

52   Abram, op. cit., p. 65.

53   Ibid., p. 62.

54   T. Ingold, *The Perception of the Environment*, Oxon, USA and Canada: Routledge, 2000, p. 9.

55   Ibid., p. 171.

56   T. Ingold, *The Perception of the Environment*, Oxon, USA and Canada: Routledge, 2000, p. 25.

57   N. Bird-David, 'Animism Revisited: Personhood, Environment, and Relational Epistemology', in G. Harvey (ed.), *Readings in Indigenous Religions*, London and New York: Continuum, 2002, p. 74.

58   D. Abram, 'The Perceptual Implications of Gaia', *The Ecologist*, 1985, 15(3), p. 100.

59   Leder, op. cit., p. 165.

60   Feminist epistemology shows that not all scientific knowing has to be disembodied (Sandra Harding, *Feminism and Methodology: Social Science Issues*, Bloomington, IN: Indiana University Press, 1987).

61   N. Bird-David, 'Animism' Revisited', op. cit., p. 96.

62   Ibid., p. 97.

63   Somatophobia refers to the fear and hostility to the body that is characteristic of Western philosophy. E. Spelman, 'Woman as Body: Ancient and Contemporary Views', *Feminist Studies*, 1982, vol. 8(1), p. 120.

64   A. Worthington, (ed.), *The Battle of the Beanfield*, Devon: Enabler Publications, 2005, p. 214.

65   Author's field notes, Jan interviewed at site A, October, 2005.

66   A. Letcher, 'Go fly a kite: A New Approach to Eco-Magick', *The Dragon Eco-Magic Journal*, 2001. Available online at: www.dragonnetwork.org/go-fly-a-kite-a-new-approach-to-eco-magick/ (accessed 21 March 2013).

67   This list is not exhaustive and there are clearly other processes of connection – erotic communion for example.

68   I have discussed the role of the wilderness effect and the threshold brook in

more depth elsewhere. See A. Harris, 'The Power of Place: Protest Site Pagans', *The European Journal of Ecopsychology*, 2011, vol. 2. Available online at: http://eje.wyrdwise.com/ojs/index.php/EJE/article/view/12 (accessed on 21 March 2013); A. Harris, 'A Life in the Woods' in G. MacLellan and S. Cross (eds) *The Wanton Green*, Oxford: Mandrake Press, 2012, pp. 68–82; A. Harris (2013). 'Embodied Eco-Paganism'. G Harvey (ed.), in *The Handbook of Contemporary Animism*. Durham: Acumen Publishing, pp. 403–15.

69 R. Greenway, 'The Wilderness Effect and Ecopsychology', in T. Roszak, M. E. Gomes and A. D. Kanner (eds) *Ecopsychology: Restoring the Earth, Healing the Mind*, San Francisco, CA: Sierra Club Books, 1995, p. 129.

70 In as far as the term 'wilderness' can be seen as a Western construct (see, for example, M. Oelschlaeger, *The Idea of Wilderness: From Prehistory to the Age of Ecology*, New Haven and London: Yale University Press, 1991), the term 'wilderness effect' is problematic. However it is widely used in ecopsychology and has become the accepted term for the specific process I'm concerned with here.

71 A. Letcher, '"Virtual Paganism" or Direct Action? The Implications of Road Protesting for Modern Paganism', *Diskus*, 2000, vol. 6. Available online at: http://web.uni-marburg.de/religionswissenschaft/journal/diskus/letcher.html (accessed 15 May 2004).

72 J. Hindle, *Nine Miles: Two Winters of Anti-Road Protest*, Brighton: Phoenix Tree Books, 2006, pp. 70–1.

73 See, for example, B. McDonald and R. Schreyer, Spiritual Benefits of Leisure: Participation and Leisure Settings', in *Benefits of Leisure*, State College, PA: Venture Publishers, 1991.

74 Greenway, op. cit., p. 128.

75 S. Shaw, *Experiential Wildness*, 2006. Available online at: www.ecopsychology.org/journal/ezine/experiential.html (accessed 21 March 2013).

76 Barry – fieldwork interview.

77 A. Letcher, 'Go fly a kite', op. cit.

78 Heathens are pagans who follow Germanic or Scandinavian traditions.

79 J. Blain, *Nine Worlds of Seid-Magic: Ecstasy and Neo-Shamanism in Northern European Paganism*, London: Routledge, 2001, pp. 61–2.

80 Greenway, op. cit., p. 132.

81 Ibid.

82 B. Patterson, *The Art of Conversation with the Genius Loci*, Somerset: Capell Bann, 2005, p. 136.

83 G. MacLellan, 'Dancing on the Edge: Shamanism in Modern Britain', in G. Harvey and C. Hardman (eds) *Paganism Today: Ancient Earth Traditions for the Twenty-First Century*, Thorsons: London, 1996, p. 147.

84 Greenwood, op. cit.

85 Bateson, op. cit., p. 468.

86 Greenwood, op. cit., p. 97.

87 Bateson, op. cit., p. 138.

88 Greenwood, op. cit., p. 94.

89 Varela, Thompson and Rosch, op. cit., p. 217.

90 Greenway, op. cit., p. 133.

91 B. Taylor, 'Resacralizing Mother Earth in the History of Earth First!', *Earth First! The Radical Environmental Journal*, 2005, vol. 6(1), p. 47.

92   Greenwood, op. cit., p. 107.
93   Jackson, op. cit., p. 328.
94   Ibid.
95   Bateson, op. cit., p. 467.
96   Although there are connections with deep ecology and transpersonal ecology that remain unexplored here, I consider eco-pagan animism to be more embodied than either. Both are part of a wider conversation and I have discussed deep ecology elsewhere (Sacred Ecology, see following endnote).
97   A. Harris, 'Sacred Ecology', in Harvey, G. and Hardman C. (eds) *Paganism Today: Ancient Earth Traditions for the Twenty-First Century*, Thorsons: London, 1996, p. 152.
98   A. Fisher, *Radical Ecopsychology: Psychology in the Service of Life*, State Albany, NY: University of New York Press, 2002, p. 103.

Interview Details

Barry interviewed in Coventry, 19/04/07

Gordon interviewed in Buxton, 29/05/07

Mark interviewed in Southend, 21/05/07

Mary interviewed in Devon, 26/05/07

Sally interviewed in London, 23/05/07

Zoe interviewed near Glastonbury, 24/05/07

Dave interviewed at site A, June 2007

Debbie interviewed at site A, October 2005

Ian interviewed at site B, June 2006

Jan interviewed at site A, October, 2005

Jo interviewed at site A, June 2007

John interviewed at site A, May 2006

Rob interviewed at Climate Change Camp, August 2006

Site A was a long-term encampment on a narrow strip of land, which was part of a local park in a suburban town in southern England. It was notable in that it included an ancient burial site. The number of full-time residents varied over the two years of my involvement from two to about 12. Although local support was strong, site A was targeted by arson and other attacks. At the time of writing site A is no longer threatened by a road-widening scheme.

Site B was a road protest in southern England established during May 2006 in several patches of woodland – some ancient – on the edge of a town. There were two camps and again numbers on-site varied, but during my visits the main camp had an average of ten residents while the second had five. Local support varied over time. The woodland is no longer under immediate threat. Sites A and B had some well-built low-impact dwellings and communal spaces.

Part 3

The legal dimension

Chapter 8

Re-embodying law
Transversal ecology and the commons

Vito De Lucia

Introduction

Modern law has had a central role in the concretization of the modern epistemology of mastery, through modern law's imbrication with capitalism, and through law's constitutive participation in what has been called the 'scientifico-legal complex', whose primary aim was that of taming and 'othering' nature.[1] Ecology, however, despite its complex ambiguities – has raised a number of challenges – simultaneously epistemological, ontological and ethical – to the way in which modern law 'thinks'. New forms of 'ecological normativity' have been identified that pose, it is argued, substantive demands on law;[2] a new 'ecological thinking' has been proposed as a new way to explore re-embodied subjectivities and re-embodying practices;[3] and new forms of critical, post-natural environmental law are being explored.[4]

Arguably, however, the largest share of attention in the emerging attempts to bring together law and ecology remains centred on two alternatives. The first embraces a human rights approach to the environmental question, and its strategy is focused on the right to a healthy environment (and other environmental rights).[5] The second focuses on how to expand the framework of subjective rights to natural entities.[6] These strategies, while promising both conceptual and concrete benefits,[7] remain largely located – with some exceptions[8] – within the dialectic of modernity.[9] As Margherita Pieraccini suggests, however, today '[t]he task for legal scholars is [to] produce a new language' so as to disentangle law and legal strategies 'from the constraints imposed by the tradition of [modernity]'.[10] This chapter aims to offer one contribution towards fulfilling this task.

Embracing a 'transversal' notion of ecology, one that exceeds its usual disciplinary and conceptual boundaries,[11] this article will try to think law ecologically without having to refer ecology or law exclusively to the modern epistemological category of nature. This transversal ecology will be linked to the practices of the commons, which embody 'insurgent' knowledges, languages and perceptions that, as we shall see, traverse and resist the trajectories of modernity and their concrete embodiment in law. Ambiguously and strategically oscillating between legal and illegal practices and spaces (yet operating within the broader space of the juridical as a social phenomenon), contemporary practices of the commons (which will form the primary referent in this article) re-embody law in the practices of bodies and communities, re-claiming the constituent power of social forces as a continuously operative manifestation of living law and embodying, I will argue, what has been called a jurisprudence of the middle.[12]

The commons, it will be argued, exceeds the modern binary space of subject/object; recuperating a meaning largely flattened and suppressed by modernity, the commons shifts the meaning of *ecological* to a multiplicity of planes not only implicating nature; and embodies the politics of epistemic location which Lorraine Code suggests is central for the needed ecological 'subversions and transformations' of the modern subject.[13] It is along these lines that the possibilities for thinking law ecologically will be explored. In what follows, I will first offer a brief account of modernity and its trajectories; then I will show how law has to a large extent concretized these trajectories, despite law being continuously immersed in a genealogical 'play of forces'.[14] Both modernity and law will be described as disembodied, and, importantly, as *performatively* disembodying. Subsequently I will discuss the concept of transversal ecology, and its implications for law. Finally I will present the perspective of the commons, before concluding with an exploration of how one can begin to think law ecologically through the re-embodying perspective of the commons.

Modernity and its trajectories

Modernity is a complex concept: '[f]ew thickets', it has in fact been noted, 'are more tangled than that in which the idea of modernity has become enmeshed'.[15] In this respect, the most I can do in this section is to outline the way in which modernity is approached in this chapter. Modernity is conceived of here as a *theoretical* category; that is, modernity is not understood (primarily) as a historical period, but rather as a mode of conceiving the world that *exceeds* its historical coordinates.[16]

This way of conceiving modernity has, I claim, two advantages. First it allows *seeing* modernity as a central dynamic in the contemporary situation. That we are in the midst of an important transition is perhaps evident.[17] However, while modernity may have exhausted its potential, it has not yet exhausted its cognitive, ideological and material influence on our contemporaneity, particularly through its interpenetration with capitalism.[18] Law in this respect has played, and still plays, a crucial role, even as some of the symbolic and material institutions of modernity seem to lose their central place in a globalized and still globalizing world.[19] At the same time, this approach also allows us to understand modernity as *one* framework, however hegemonic, whose genealogy is deeply entangled with all those 'counter-modernities'[20] that it has subjugated and marginalized, but not erased. Modernity then arguably remains the dominant worldview even at a juncture where its main categories are in a state of crisis, and have been subjected to a multiplicity of critiques.

The trajectories of modernity

A useful way to conceptualize the processes that underpin modernity in a dynamic manner and to remain consistent with the idea that modernity exceeds its historical coordinates is to deploy the concept of *trajectory*. Unlike the term 'break' that, often used to describe the passage to the modern era,[21] speaks of a neat separation between a 'before' and an 'after' localizable in a mythic origin,[22] the concept of trajectory captures the plurality of discontinuities *and* the diversity of spatial and temporal processes that brought forth and still sustain modernity. Never quite complete or total, a trajectory remains open (and to some extent also vulnerable) to the contestations, deflections, corrections and adjustments implicit in the genealogical 'play of forces'[23] underlying conceptual developments and social practices alike. In other words, trajectories are dynamically inscribed in the hegemonic relations of power and resistances sustaining modernity.[24]

For the purposes of this article (and for reasons of space) the plurality of relevant trajectories that can be in principle identified will be gathered around a single central meta-trajectory – *disembodiment*. In this meta-sense disembodiment represents a set of abstracting processes, premised on the transcendental obsessions of *one particular, hegemonic strand* of western philosophy.[25]

In a first sense, disembodiment can be understood as a process (the social dynamic aspect) and conceptual framework (the static, ideological referent) that constructs the basic categories of the modern binary worldview. The Cartesian 'conceptual body excision'[26] is in this respect decisive, as it locates subjectivity and autonomous agency only in the rational mind. The extended world of matter then (including the body within which the mind resides)[27] is re-conceptualized as devoid of reason, affectivity and agency, and governed by immutable laws.[28]

Here the crucial thresholds of modernity are established, the ones separating mind from body; subject from object; culture from nature; reason from emotions. Nature can then become a methodological device instrumental to the political epistemology of modernity, and part of a 'fabulous ploy' aimed at deciding 'who will be allowed to talk about what, and which types of beings will remain silent'.[29]

This conceptual manoeuvre provided crucial *ideo-ontological*[30] resources for the enactment and implementation of the modern epistemological project of *mastery* and of the economic project of capitalism. Through the mediation of disembodied rational cognition[31] the world undergoes a process of simplification through mathematical reason and a reductionist method.[32] Nature, whose new modern conceptualization 'contained manipulability at its theoretical core',[33] becomes homogenous; it is everywhere the same: 'be it heaven or earth', it is bound to obey the same laws.[34] Thus the new category of nature is re-organized as an assemblage of fungible parts, no longer bound to a specific place. These trajectories mark then the 'fundamental tendency toward abstraction' inherent to modernity and instrumental to capitalism.[35]

Disembodied and disembodying law

How does law fit into this picture? Anna Grear suggests that the 'heart of the problem of legal disembodiment lies deep in the philosophical foundations of liberal legal rationality'.[36] In this respect the modern trajectories just outlined are arguably *objectified* in/as law.[37] However it is important to keep distinct the ideal or ideological terrain from the social terrain where such ideas and ideologies operate and produce concrete effects.

Operating in both terrains, law invites contestations. As a rational, disembodied discourse, modern law reflects its ideo-ontological commitments. As a social process however, law is a privileged point of emergence of conflicts.[38] As such law is one of the central sites where hegemony is asserted, negotiated and reproduced.[39] These trajectories then are objectified in/as law in the (more limited, nuanced) sense that law is reflective of a particular hegemonic configuration that changes over time, but remains dominated by a particular operating code.[40]

Importantly, these trajectories arguably still underpin a disembodied construction of the subject; still underpin the construction of nature as separate from culture; still underpin the transformation of the 'vulnerable living order'[41] into ecosystem services and biological and genetic resources, calculable, measurable and reducible to their instrumental value through the operations of science, markets, technology and, importantly, law.[42] These trajectories maintain a hegemonic grip *on*, *through* and *as* law. From this point of view law appears in two distinct, but crucially related ways: it is disembodied; and it is disembody*ing*.[43]

Modern law, it has been observed, has 'systematically suppressed' the inevitable material embodiment of human beings.[44] As a rational discourse, modern law is reflective of, and enforces in complex and frequently indirect ways, Cartesian disembodiment. The body, like nature, is subjected to mechanical laws, and is consequently bound and heteronomous, in opposition to the rational mind, which is free and autonomous. The subject furthermore, characterized by abstract and universal characteristics, transcends the particularities of embodiment, while the body is excluded,[45] and is turned into a 'tacit threat'[46] to the autonomy and independence of the rational subject, whose self-recognition as subject requires only the capacity to think. In this sense then the disembodiment of the modern subject consummated with Descartes has been translated through a series of intellectual manoeuvres (in fact some long preceding Descartes)[47] into the abstract, rational, liberal, disembodied legal subject.[48] This subject is a rights bearer by virtue of its[49] rational agency.

The legal subject is in this respect exhaustively located in the rational mind, whose productive force is a rational will that creates law.[50] Law itself is construed as a disembodied set of general and abstract rules inhabiting a rational universe detached from the concrete particularities of lived experience. Obediently, if sometimes reluctantly,[51] responding to the demands of capitalism, law takes the commodity form.[52]

This conceptual and ideological disembodiment however remains effectively only a quasi-disembodiment, as Anna Grear points out,[53] given the irreducible presence of the biological body, which makes its ideo-ontological exclusion persistently problematic.[54] Consequently, the universal legal subject translates into a template of exclusion with regards to both other humans – those not approximating well enough its particularistic template[55] – and the broader non-human world, relegated into the role of harvestable resource. Moreover, while remaining conceptually disembodied, the subject operates concretely on the world, in part through (private) property, which is, paradoxically, one of the prime juridical sites for the embodiment of the disembodied will of the subject.[56]

And it is at this crucial intersection of the subject with property (in its double sense of private and public/State property) that law, descriptively disembodied, becomes *performatively disembodying*. Law produces and shapes the world in particular ways, with very deep and material consequences,[57] and modern law predominantly 'thinks' in terms of private property rights, following an 'epistemic imaginary' operative at least since Locke:[58] land and nature are automatically conceived as consisting of discrete parcels that can be owned,[59] or as a repertoire of fungible entities, available to the subject as its property, and exchanged through contract.[60] At the other side of what Mattei calls the 'pincer' of modernity,[61] but following the very same epistemic imaginary, is the public form of property, whereby a State owns its territory.

In a particular historical juncture such as the contemporary one however, where multiple ecological crises overlap and dramatically reinforce one another, it is crucial for 'legal scholars [to] produce a new language'[62] that radically disarticulates the disembodied and disembodying trajectories of (legal) modernity. The next section will accordingly discuss the role of ecology in relation to this task, in order to begin exploring the possibilities for a re-embodied law, and for a re-embodying legal philosophy.

Transversal ecological thinking

Ecology is one of the central questions of the postmodern juncture, both in relation to the multiple and overlapping ecological crises; and in relation to its capacity to fundamentally problematize the categories of (legal) modernity.[63] Ecology however raises complex questions. Due to its double epistemic role – as a science and as a normative framework – ecology is fundamentally affected by a 'moral ambivalence'.[64] Moreover ecology is affected by an irreducible genealogical complexity,[65] and lies at the centre of discursive contestations aligned with a variety of political projects, responding to conflicting paradigmatic orientations, and aimed at imposing hegemonic closures.[66] If ecology is in fact used to derive philosophical implications aimed at challenging the modern construction of nature, it is also deployed as a legitimating framework for the continued enforcement of nature as a modern category, and for its exploitation.[67] If the discourse of ecology inspires an apparent radicalization of legal theory,[68] it also enables new biopolitical regimes aimed at the regularization of the provisions of ecosystems goods and services.[69] What ecology then? Here I will take on board recent expansions of the scope of ecology,[70] in order to mobilize its transversal and re-embodying potential.

Guattari suggests for example that ecology requires us to think 'transversally',[71] that is, we must simultaneously embrace the natural, technical, social and psychological plane, if there is to be hope of addressing the 'increasing deterioration' of each.[72] Lorraine Code similarly suggests that ecological thinking, by subverting the neat separation of the personal, the social, and the natural, allows the formation of transversal links between, say, ecosystems and international capitalism.[73] Such perspectives resonate with Latour's revelation of nature as a methodological device instrumental to the political epistemology of modernity, whose primary function is that of producing a threshold in relation to which voice can legitimately be heard.[74] The displacement of the modern epistemological category of nature however does not displace nature as *world*.[75] So if this reading of ecology dismantles the thresholds of modernity, it does not dismiss the complex, embodied presence of the living world.[76]

Ecology then should be understood to entail a plurality of registers engulfing the modern binaries in an irreducible, transversal wholeness. This engulfment overwhelms and then displaces the epistemological paradigm of modernity through which independent and autonomous subjects know only isolated objects.[77] Understood in this sense, ecology likewise displaces the ideo-ontological foundation of modernity. It is only on these premises, I argue, that ecology can be mobilized towards the elaboration of a critical legal framework responsive to the demands set out in the introduction: imagining a new (legal) language not reducible to modernity. A transversal ecology then is neither anthropocentric nor ecocentric; it does not operate from any centre.[78]

The next sections will first briefly outline how an ecological matrix of perception re-imagines subjects and objects; and then how ecological knowledge is inevitably unstable, negotiated, political and situated.

Ecological forms

An ecological understanding of the world suggests a very different onto-logical commitment than that embedded in modern law. This difference takes two immediate forms. First, each ecological entity is *situated*, that is, is constitutively immersed in a network of relations. Second, subjectivity as a result – as Code and Guattari similarly suggest, even if by way of different arguments and conceptual pathways – can be attached to a singularity, that is, a provisional embodiment of a multiplicity of relations enmeshed in a whole as a part, but no longer to an individual understood in the modern sense of an undivided (and hence undividable) unit. Indeed, what modern philosophy conceive of as individuals, may be merely 'terminals' of a multiplicity of interacting 'components of subjec-tification';[79] and it is at the crossroads of these interactions and conflicts that the interiority of an individual (or rather, of a singularity) is established.[80]

This relational commitment of ecological thinking has important effects on law, since legal notions, while conveying 'a particular partitioning of the world',[81] are dependent on particular 'perceptions and representations of reality'.[82] A shift in the perception of reality translates (perhaps inevitably) into a modification of the cognitive premises of law.[83] Furthermore, a range of new elements and entities may become 'ecologically relevant',[84] upwards, downwards and laterally with respect to 'standard' individual subjects and objects, and require a novel and appropriate legal translation. Importantly, and we will return to this point, the relevance of these ecological entities is not premised on their essential being, but on their being situated in relation to particular questions, so that the set of ecological subjectivities may shift in space and time.[85]

This ecological 'shiftiness' is however a 'source of problems' for law,[86] because the ecological relational ontology cannot be appropriately trans-lated into a legal ontology premised on a 'sharp separation between subject and object',[87] and where both maintain clearly delineated boundaries.[88] A relational and systemic ontological matrix invites instead an appreciation of the ways in which singular entities are constitutively situated within complex ecologies as complementary parts within wholes.

Uncertain knowledge and epistemic location

A central insight of ecology relates to the instability of knowledge. Ignor-ance and uncertainty acquire a crucial epistemic and normative role.[89] This endemic (and perhaps insuperable) uncertainty deriving from the complexity of non-linear and cross-scalar ecological processes leads to what has been called 'truth pluralism'.[90] Every decision then carries with/in it specific normative, ethical and political commitments arising from both scientific and legal processes,[91] as knowledge and values are entangled on both objective and subjective grounds.[92] This situation has prompted scholars to describe law, and environmental law more specifically, as post-modern,[93] hot,[94] irreducibly mired in genealogical tensions and situated between competing narratives.[95]

If knowledge is uncertain, unstable and negotiated, it becomes important, following Latour, to embrace a shift in focus from matters of fact to matters of concern.[96] And in the context of this shift law, no longer defensible as a *technical* domain, must exploit the productive potential of epistemic instability, and accept 'the responsibility to solve problems which science cannot decide and that are linked to uncertain outcomes'.[97] It will be important then, as Lorraine Code suggests, to emphasize the centrality of 'epistemic location' in relation to the creation, negotiation and circulation of ecological knowledge – 'down on the ground', as Code aptly puts it.[98] In this sense 'the nature and conditions of the particular 'ground', the situations and circumstances of specific knowers, their interde-pendence and their negotiations' all become relevant, even crucial, factors.[99] This points in the direction of a jurisprudence of the middle rather than of the centre. And, as an instantiation of a situated practice of the 'middle', it is precisely in the particular 'grounds' that the commons know and produce law.

The perspective of the commons

The commons, once the dominating form of resource governance, were effectively destroyed at the onset of modernity.[100] Some marginal enclaves still endure and resist, but as nothing more, it has been noted, than 'the institutional debris of societal arrangements that somehow fall outside of modernity'.[101]

Indeed it has been observed how to legal modernity the commons are 'invisible', or worse, 'unthinkable'.[102] As a dis-homogenous plurality of collective forms of properties,[103] the commons are a 'monstrosity'[104] disruptive of the civilized (legal and economic) order of the State,[105] because they exist outside of the processes of the capitalist circulation of com-modities.[106] Yet a number of social practices (which, as we shall see, are *ecological* in the transversal sense outlined above) are re-claiming the commons both materially as common goods or resources, and more broadly as a philosophical, political and juridical horizon.[107] Indeed, the genealogy of traditional subsistence commons is currently converging with contem-porary political and legal articulations of the new commons, such as 'civic',[108] 'digital'[109] or 'cultural' commons,[110] (often collectively called 'immaterial' commons)[111] and are parts of a process of shifting legalities and inter-legalities beginning to occupy spaces at the interstices of modernity, exploiting fractures and moving across porous cultural and conceptual territories,[112] perhaps also possibly coalescing into insurgent trajectories.[113]

A comprehensive account of the commons is beyond the scope of this article.[114] It will be however useful to briefly locate the central coordinates of the commons, before drawing lines of connectivity with ecology and offer a legal philosophical reading.

Beyond taxonomic closure

Given the diversity of the practices of the commons, attempts at providing some taxonomic clarity have recently emerged. Marella has for example tentatively organized the commons in five categories,[115] with the aim of overcoming their 'problematic heterogeneity',[116] which hinders, Marella suggests, an appropriate juridical construction of the category,[117] to the extent that a too wide usage may compromise the 'expressive efficacy' of the concept.[118] Marella then aims at organizing this heterogeneous, composite, moving and at times contradictory social experience, in order to capture it legally through a set of taxonomical closures that *fix* the commons in relation to determinate and shared properties.[119] Interestingly however (and we shall come back to this point later), the first shared character that Marella identifies is negative: that no common legal regime can be identified and attached to the commons. Their diversity makes subsumption under a neat taxonomy impossible,[120] and requires that legal strategies be identified and deployed only with regards to the particular case.[121]

Weston and Bollier have also attempted to capture the diversity of the commons through a set of categories, or typologies.[122] But they too conclude that the commons cannot be subsumed under a 'universal template', since each particular commons 'is grounded in particular, historically rooted, local circumstances'[123] And this is, indeed, a *central* characteristic of the commons, to which we will shall shortly return. In this respect, the commons follow a conceptual and juridical logic of their own and are inscribed in a metaphorical horizon utterly alien to modernity.

Genealogical and juridical horizon of the commons

The commons, as we have seen, resists and eludes taxonomic closures (taxonomy being a thoroughly modern operative mechanism).[124] At a general level, the commons can be understood as 'a matrix of perceptions and discourses'[125] following conceptual, philosophical and juridical trajectories incommensurable with modernity. Such a matrix, as Weston and

Bollier suggest, 'can loosely unify diverse fields of action now seen as largely isolated from one another'.[126] And indeed the commons today[127] share genealogy, epistemic imaginary and conceptual and operational mechanics with both traditional, pre-modern European commons and with all those subsistence commons still operating at the material and conceptual periphery of the modern, capitalist world. In this light, the perspective of the commons can be understood as an 'insurrection' of what Foucault calls 'subjugated knowledges'.[128]

In a first sense, subjugated knowledges are 'historical contents that have been buried or masked in functional coherences or formal systematizations'.[129] These historical contents, retrievable through scholarship, are crucial insofar as they alone enable 'us to see the dividing lines in the confrontations and struggles that functional arrangements or systematic organizations are designed to mask'.[130] In a second, quite different sense, subjugated knowledges comprise 'a whole series of knowledges that have been disqualified as nonconceptual [...] insufficiently elaborated [...] naïve [...] hierarchically inferior,'[131] to the extent that they do not bear the mark of a scientific discourse. And while the distance between the two forms is in principle significant (the first relying on scholarship and the second on local, disqualified knowledges), Foucault emphasizes how their parallel re-activation makes critique possible.

The commons arguably embodies this parallel re-activation: through their forms of practice and modes of collaborative and transversal research, the commons traverse the social and the scholarly, the militant and the conceptual, conjoining this double re-activation in insurgent practices.[132] Shifting the discourse from knowledge to law, this parallel insurrection re-activates local legalities, of which the commons are a historical and contemporary embodiment. In this respect these insurrections are most importantly insurrections against those processes of centralization that organize knowledge, and – most crucially for the purposes of this chapter – modern law, as a form of power legitimated by a (legal) scientific discourse and as a form of technical knowledge.[133] In so doing the commons re-activate a genealogy which Linebaugh, in his account of historical commons,[134] maps to three central characters that the commons, as a juridical horizon, share.

First, commoning is 'embedded in a labor process'.[135] The commons then is an activity, a dynamic process, a *doing* that is constantly *becoming*. In this crucial respect common rights are different from human rights,[136] as the former are 'entered into by labor'.[137] This is in full agreement with the etymology of the word 'common', comprising the Latin words *cum* and *munus*, meaning 'with a burden, or duty', particularly in the sense of co-obligation.[138] But implicit also is the requirement to care for the commons as a condition for participating in its governance and to benefit from the stream of utilities it may provide. This implicates an ambivalent, complementary fusion of both duty and right, within a common horizon of participation: each commoner has either both, or none.[139] There is no disembodied, abstract subject as the bearer of (disembedded) rights here; both legal subjectivity and legal rights are (re-)embodied through practice.[140]

It is also evident that the commons, while singularly always situated in a particular ground, share the same juridical horizon. In fact, if Linebaugh speaks of historical commons, participation and 'labouring' are still constitutive dimensions in new, contemporary practices. The Statute of the Teatro Valle of Rome (a symbol and catalyst for the entire commons movement in Italy)[141] offers in this sense an illustrative example. Participation in the Foundation constitutes 'a personal commitment to the collective political action' of the theatre as a commons (article 5(1)). Moreover, while anyone can become a commoner (after however having participated as observer for at least three assemblies), their status as a commoner can only be maintained through the voluntary enactment of 'practices of care' (whether physical or intellectual labour), and by participation in commons governance (article 7(2)). Finally, and more specifically, all commoners have the 'right-duty' (joint word in the original: '*diritto-dovere*') to participate to the management of the commons (article 7(4)).

A second element that Linebaugh identifies is the collective nature of commoning:[142] there is no individual, autonomous subject at work here, but a complex relational whole, where singular parts are embedded in a larger whole. The commons, in other words, always refers to a community of commoners. But, importantly, a commons is always a complex assemblage of peoples, places and times, so that a commons is always constitutively linked, through practice, to a particular 'ground'.

Finally, commoning exists within its own temporality, ground 'deep into human history'.[143] As such it is independent of the law of central/izing institutions (such as the State),[144] and exists within its own juridical horizon, a horizon particular to a community, a place, a particular 'ground', to recall again Code's ecological terminology. This latter character – the location of the commons outside or beyond (or even against) the legal order of the State – is of paramount importance, as we shall see later.

Commons as practices

Connected to their resistance to taxonomic closures, the commons, particularly in contemporary political articulations, rejects any essentialism and cannot be framed as something already there. This is true even in relation to natural resources, often framed *objectively* as commons.[145] The commons is, instead, an open category. As Mattei emphasizes, the commons is discovered, affirmed and re/produced through social practices and/or struggles, and its continued existence is contingent upon the continued practices which affirm and (re-)claim it:[146] indeed, the very social conflict that leads to the recognition of something as a commons is considered an integral part of that commons.[147]

In this sense, and crucially, the commons responds to a language of *doing* and *becoming*, rather than to a language of *having*.[148] The key legal question is not linked to title, but to access.[149] A commons observes Rota in this respect, does not, strictly speaking, belong; it is rather the commoners that identify themselves with the commons.[150] The commons, accordingly, is not primarily an object of (proprietary) rights (which in turn requires a subject), but is rather a *relation* and a *practice*. Practice, explains Barcellona, is a stance taken before the world, an embodied perspective that transforms the world and structures desires, intentions, calculations and decisions.[151] Practice is thus performative, and in this respect can be disembodying or re-embodying. Modern practice (re-)structures the world so that the world can fit modernity's disembodied ideo-ontology, thus concretizing the trajectories of modernity such that humans are restructured into abstract disembodied rights bearers, and nature into an ensemble of fungible commodities. Then the practices of the commons are, as both Grossi and Linebaugh emphasize,[152] inimical to the commodity form that turns every relation into an abstract, disembedded exchange relation.[153] It is in this respect that the commons, then, are re-embodied and re-embodying practices.

The notion of practice also has important epistemic implications. The experience of the commons in fact, suggest Hardt and Negri, is able to break 'the epistemological impasse' of modernity, by 'cutting across' the modern binary opposing universal and particular.[154] More specifically, Hardt and Negri, following Wittgenstein, locate a pluralized truth in the midst of language games understood as forms of life. Truth is thus removed from any transcendental plane, and re-located 'on the fluid, changeable terrain of practice, shifting the terms of the discussion from knowing to *doing*'.[155] This resonates with Code's notion of 'truth to',[156] which is a form of interpretation (rather than of verification) that is 'textured and responsive',[157] as well as 'responsible [to] local sensitivity'.[158] Ultimately, truth is a form of life responsive to the embodied world, and a form of responsible knowledge, meaning a form of knowledge cognizant of the 'multiply contestable' nature of categories and taxonomies that impose permanent closure on the living world.[159]

Doing also crucially implicates a responsibility immanent in the doing, to the extent that doing entails immediate consequences on an embodied plane, as opposed to *disembodied thinking* that, abstracted and rarefied, is performatively disembodying through the conceptual and material displacement of its consequences to another place and another time.[160]

Transversal themes for a re-embodied and re-embodying legal philosophy

The preceding discussion has shown how the commons capture crucial aspects of transversal ecological thinking, and embodies trajectories incommensurable with those of modernity. It is now possible to suggest some initial ways in which the perspective of the commons and transversal ecology can help to re-embody law.

It is perhaps useful to recall how some of the major questions in legal philosophy traditionally relate to the concept of law; the theory of sources; and the theory of the legal subject. This section will touch upon each of those questions. Of many possible, I will however only touch upon *some* overlapping themes. I discuss them separately and under different headings but they all speak the same language – a language that can be described as a language of re-embodiments – and offer each one particular nuance of the same whole.

The first theme relates to the central symbolic and material function of the body. A second theme relates to an institutional concept of law, and, relatedly, to a radical legal pluralism. The third theme relates to the commons as embodying an understanding of law as immersed in life, rather than subsuming life within its categories, in this respect *re*-embodying law in a manner deeply resonating with what Philippopoulos-Mihalopoulos calls a jurisprudence of the middle. Finally, both the commons and transversal ecology suggest that *any ecological* philosophy of law must crucially focus on the *content* and *effect* of law. The question to ask then is no longer 'is law valid?' nor is it 'is law efficient?'[161] The relevant question is rather whether law is, with Code, responsive and responsible. Of course it is no longer possible to find a universal reference against which to measure law; epistemic location and the transversally ecological practices of the commons however alerts us to the possibility of negotiating responsive and responsible law *in the grounds*. A further, transversal, theme is, finally, the immanence of conflict, whose primary implication is that there is no neutral law, and that legal scholars cannot find solace in an alleged neutrality of scholarship. One must take sides.

The body

The body assumes a central symbolic function, insofar as it is juxtaposed to the modern disembodied philosophy of the subject, and contests the thresholds that modernity establishes between biology and history; mind and matter; reason and passion; harmony and conflict; civilization and the state of nature.[162] However, the body is no longer a category responding to the Cartesian ideo-ontology. It is, rather, a *somatic place*, a 'cultural body' [...] amenable neither to ideal nor [material] reduction'.[163] In this respect, the body reclaims a central cognitive function, to the extent that, as Guattari noted, thinking is not sufficient in order to be, since 'all sorts of other ways of existing have established themselves outside consciousness'.[164] This resonates with Bruno's notion of the civilizing hand, that is, of the cultural and technical primacy of the hand with respect to the mind.[165] In a crucial sense then the body as a somatic place is not merely the material substrate of reality for which re-embodiments intend to re-claim central place in a merely *inverting* manoeuver that would as such fail to dislodge the modern obsession for the centre.[166]

The body as a somatic place then, I argue, is *aesthetic, political* and *ecological*. It is aesthetic in the crucial sense that, as the etymological semantic field of the word shows, the body encompasses simultaneously sense perception (importantly including what Anna Grear calls the 'non-cognitive porosity' of the body),[167] intellectual discernment and moral cognition.[168] These modes of knowing are diffuse and follow different logics that are complementary. The body in other words is epistemically plural. The body is political because it enacts embodied practices of subjectification (in the active sense of asserting one's presence), and is simultaneously the object of forms of power aimed at its normalization, disciplining, productive enhancement and subjectification (in the passive sense of *assoggettamento*). It is, in other words, immersed in conflict. The body is ecological as it is inevitably situated in an open and transversal ecology that the body, simultaneously biological, psychological, technical and political, traverses.

Living law, insurgent critique

As we have seen, the perspective of the commons, with its complexity and diversity, cannot be accommodated within the confines of modern homogeneous simplicity and exclusionary binaries. As a complex and plural ensemble of social practices, the commons cannot be described, captured, normalized or classified through abstractions and general properties. The notion of insurrection, briefly discussed above, very aptly describes the re-claimed role of disqualified legalities subjugated by the hegemonic legal order of the State. The plural practices of the commons and the epistemic location of transversal ecology, particularly in relation to their insurgency, resonate with a theory of law known as institutionalism, as law is (re-)located – indeed, (re-)embodied – in organized social practices, or what in legal theoretical terms are called simply *institutions*.[169] Here I refer in particular to the notion of institution elaborated by Italian legal philosopher Santi Romano.

Romano describes an institution very widely as 'every social entity or body' that has a significant enough measure of stable pattern, form and/or organization.[170] Romano thus radically identifies the social and the legal (so that neither has causal, logical or temporal priority). A legal order,[171] from this perspective, 'is the concretisation of a social fact [...] the effectiveness of its structure'.[172] Romano however tended to close his institutional theory around the all-encompassing institution of the State, integrating all others within a hierarchical structure.[173] The commons on the other hand point to a much more radical orientation:[174] social practices are always already legal practices,[175] and law becomes *living* law. Any order including competing social forces and legal claims from this perspective is not subsumed within a higher structure (such as the State). It takes rather the form of a complex and complementary assemblage constituted through this conflict, and always ambiguously balancing and re-balancing. The architectural model for this radical institutionalism is not the Constitutional State, but a bottom-up, rhizomatic federalism, without a centre and without a top.[176]

This institutional perspective further points in the direction of a radical legal pluralism.[177] If the social is always already juridical, and each social institution carries and re-produces its particular legality, the world (what Deleuze and Guattari would call the plane of immanence) is then criss-crossed by overlapping juridical institutions carving their own constituent space through the production of a plurality of ecologically situated legal habits, a multiplicity of legalities co-extensive with the social institutions that form, self-recognize and act in the world. In this respect, the commons are practices and institutions hooked onto particular features of the world, to which they claim access through their practices, and in relation to which they claim responsibility and juridical competence.

Law in the middle of life

Transversal ecology and the perspective of the commons both speak of an immersion in life. And this immersion begins in the middle of that 'plane of immanence' that contains without transcendental excess 'its own origin, causality and teleology'.[178] 'The middle', suggests Philippopoulos-Mihalopoulos, 'does not allow for a perspective that calls itself an origin'.[179] Thus located in the middle, in the midst of a sensuous and unpredictable life, the commons as a legal practice disarticulates the entire modern logic of presupposition.[180] The perspective of the middle is able to disarticulate the modern obsession with the centre; an obsession represented, within the context of environmental philosophy and law, by the binary anthro-pocentrism/ecocentrism,[181] and that condemns legal philosophical reflections to oscillate between what Hasley calls the sacred and abject views of nature.[182]

Through the situated doing of the commons on the other hand, the world is neither sacred nor abject, and law is produced in the middle as a form of life. Yet the middle does not offer any guarantee of being a *good* middle (and from now on middle, plane of immanence and world will be used interchangeably). Indeed, as Philippopoulos-Mihalopoulos warns, the world is complex, and its space is 'a space of struggle'.[183] This is an important point, as one of the crucial thresholds of modernity is that between order and chaos (in all its declinations: peace and conflict; culture and nature; civilization and savagery, etc.). Yet life, being in/the middle, is necessarily traversed by conflict. This is the premise of an entire philosophical[184] and legal[185] tradition that resonates deeply with contemporary articulations of the commons.[186] Conflict, in this tradition is not inimical to order: it does not precede it; nor does it follow it. Conflict, through the ambiguous productive potential it generates, is constitutive of order (an order however always provisional).[187]

Displacing the disembodied subject through relations of complementarity

The commons is fundamentally a set of practices of social cooperation in relation to a set of common goods understood or (re-)claimed as commons. Moreover, the commons traverses the boundaries that modernity establishes between nature and culture, as commons are at once natural, social, technical and psychological. One of the main operative concepts in this respect is complementarity. Following Heidegger, the meaning of complementarity will be explored through etymology, in order to 'open it up', and show its broad semantic field.[188]

Complementarity has its root in the Latin *complere*, which indicates a literal (filling up some empty vessel) or metaphorical (carrying through some action) completion of something.[189] The semantic field of complementarity further indicates the relation of two (or more) parts that together form a whole,[190] without any particular relation of subordination permanently or univocally implicated by their interaction; in other words, without a centre. The part–whole relationship implies moreover that the complementary constitutive parts of the whole cannot be reduced only to rational subjects, but *include* the people, the lands and the entire living world of that particular ground where a commons is established, with all the reciprocal normative demands that this complex assemblage entails,[191] and inclusive of that immanent conflict upon which the productive tension of a responsive and responsible equilibrium can be maintained. In this respect the idea of complementarity also captures the fact that transversal ecological relations are qualitative, asymmetric and ambiguous, rather than quantitative, neat and equal. The invincible disembodied subject evaporates, and in its place one finds vulnerable complementarities.[192]

In this respect, complementarity is incommensurable with the modern primacy of (abstract) equality. Equality, enabling autonomous, free and equal owners of commodities to interact through market exchanges,[193] underpins the key legal template of capitalist modernity, that is, the 'abstract and general individual' possessor of individual rights and free producer of law.[194] Through this template – a template that has facilitated the corporate capture of human rights[195] – the world is understood and categorized in terms of subjects (owners and sovereigns) and objects (owned).

Complementarity offers the opportunity to re-think law along a different, re-embodying trajectory. This resonates with Lon Fuller's idea of law as a 'language of interaction'.[196] Particularly in its most important character of social practice (as opposed to 'officially declared or enacted law'),[197] law is understood by Fuller as inevitably and essentially inter-active.[198] Such law 'develops out of human interaction',[199] as the practical result of 'complementary expectations'.[200] Law is in this sense 'a program for living together'.[201] Law however must be able to respond to the complex reality of plural and shifting subjectivities that both ecology and the commons enable and make visible.

Accommodating shifting ecological subjectivities

As we have seen, the new, mobile ontological commitments that ecology confronts us with, pose problems to modern law, whose grammar and operational mechanics is dependent on clearly delineated subjects and objects and on linear causal relations. As Christopher Stone underlined in his seminal article 'Should trees have standing?',

> 'there are large problems involved in defining the boundaries of 'natural objects' [...] from time to time one will wish to speak of that portion of a river that runs through a recognized jurisdiction; at other times one may be concerned with the entire river; or the hydrologic cycle – or the whole of nature. One's ontological choices will have a strong influence on the shape of the legal system'.[202]

Indeed, ecosystems are generally considered to inhabit shifting ontologies, depending on the particular perspective, point of view or problem being confronted with. How can law then accommodate the shifting subjectivities that ecology confronts us with, and give voice to the pervasive and embodied difference of the 'vulnerable living order'?[203] Widening the scope of the subject as a vehicle for giving a legitimate legal voice to natural and other non-human entities may be practical in order to immediately exploit the opportunities existing legal systems afford.[204] However, one may suspect that (besides the unresolved problems with respect to *shifting* subjectivities) the template of the subject will always tacitly operationalize its exclusionary orientation, by way of inducing ontological and epistemological transformations and simplifications in order for the subsumption of the novel subjectivities to become legally viable.[205]

What then? A jurisprudence of the middle may offer some help, to the extent that, located 'in the middle [...] of concrete situations',[206] it is able to operate, as we have already mentioned, 'without preconceived structures'[207] and without fixed categories of subject and object. How to operationalize this is a question that cannot be addressed here. However it is clear that a re-embodied concept of law must be responsive to the immanent, situated normativity of the world. Thus classic concepts such as 'the nature of things' may regain some centrality, albeit within a framework of situated and negotiable knowledge that requires law to be responsive, responsible and, most importantly, performatively re-embodying.

In this respect, the vulnerability of bodies is an important consideration that becomes legally operative (in a performatively re-embodying sense) to the extent that it seeks to disarticulate a legal rationality that has granted 'legal humanity'[208] to abstract, disembodied entities such as the corporation, while denying access to legal subjectivity – with all the rights and privileges it affords – to a large ensemble of embodied, vulnerable entities.[209] This legal rationality validates one analogy (rational subject > disembodied legal fiction) and discards another (embodied vulnerable human > embodied vulnerable non-human entity),[210] providing further evidence of how the ontological possibilities afforded by the Cartesian disembodiment have been mobilized to sustain a particular articulation of modernity imbricated with capitalism.[211]

Concluding remarks

As we have seen, a disembodied world is devoid of independent meaning, reducible to its material component parts, calculable and susceptible to manipulation. To be sure, the world is there just as before, yet it is suddenly silent, passive and devoid of normativity. Paradoxically, through this disembodiment, it becomes *materialized.* Disembodiment is thus not about (or not only about) dephysicalization, but is (most importantly) a *strategy and praxis* of negation of meaning to that which is outside of the autonomous subject, the sovereign centre of the modern worldview. Law, we have seen, has functioned as a crucial mechanism embodying and concretizing the trajectories of modernity.

This chapter has tried to take a step in the direction set out in the introduction, by way of tracing lines of connectivity between a transversal concept of ecology and the practices of the commons, hence combining the thinking of theory with the doing of practices, in order to open space for articulating an ecological, re-embodying legal philosophy.

Notes

1 J. Holder, 'New Age: Rediscovering Natural Law' *Current Legal Problems*, 53, 2000, p. 165

2 M. Tallacchini, *Diritto per la Natura. Ecologia e Filosofia del Diritto*, Torino: Giappichelli Editore, 1996

3 L. Code *Ecological Thinking: The Politics of Epistemic Location*, Oxford: Oxford University Press, 2006

4 A. Philippopoulos-Mihalopoulos, *Law and Ecology: New Environmental Foundations*, London: Routledge, 2012

5 M. Pieraccini, 'Reflections on the Relationship between Environmental Regulation, Human Rights and beyond – with Heidegger' in A. Grear and E. Grant (eds) *Thought, Law, Action and Rights in the Age of Environmental Crisis*, Cheltenham: Edward Elgar, 2015, forthcoming. For a detailed treatment of its status and of its different elements (procedural, substantive, etc.) see B. Weston and D. Bollier, *Green Governance. Ecological Survival, Human Rights, and the Law of the Commons*, Cambridge: Cambridge University Press, 2013, especially chapter 2

6 Thus the seminal article of Christopher Stone; Thus the large majority of the scholarship contributing to the development of Earth Jurisprudence; for some examples see C. Cullinan, *Wild Law: A Manifesto for Earth Justice*, Cape Town: Siber Ink, 2002; P. Burdon, *Earth Jurisprudence: Private Property And Earth Community*, London: Routledge, 2015. For some exceptions see, e.g. M. Tallacchini, 'A Legal Framework from Ecology' Biodiversity and Conservation, 9:8, 2000, as well as G. Winter, 'Ecological Proportionality – an Emerging Principle of Law for Nature?' and J. Laitos, 'Rules of Law for Nature's Use and Nonuse' both in C. Voigt (ed.) *Rule of Law for Nature. New Dimensions and Ideas in Environmental Law*, Cambridge: Cambridge University Press, 2013

7 See e.g. the Constitution of Ecuador, particularly article 71, which states that 'Nature [...] has the right to exist, persist, maintain and regenerate its vital cycles, structure, functions and evolutionary processes', which has to date given rise to at least one court case, see decision of Provincial Court of Justice of Loja, No. 11121-2011-0010, March 30, 2011, and, for an analysis, S. Suárez, *Defending Nature: Challenges and Obstacles in Defending the Rights of Nature Case Study of the Vilcabamba River*, Bonn: Bibliothek der Friedrich-Ebert-Stiftung, 2013. But rights of nature legislation has also been passed in Bolivia. For one of the most recent reflections on rights of nature see, e.g. A. Grear, *Should Trees Have Standing: 40 Years On*, Cheltenham: Edward Elgar, 2012; for a now classic overview (though not specifically legal) see R. Nash, *The Rights of Nature: A History of Environmental Ethics*, Madison, WI: University of Winsconsin Press, 1989

8 A more nuanced and more sophisticated approach is for example offered in A. Schillmoller and A. Pelizzon, 'Mapping the Terrain of Earth Jurisprudence: Landscape, Thresholds and Horizons' *Environmental and Earth Law Journal*, 2013;3: p. 1

9 For a discussion of how modernity maintains resistances within its dialectic see e.g. M. Hardt and A. Negri, *Commonwealth*, Cambridge, MA: Harvard University Press, 2011

10   Pieraccini op. cit., p. 17

11   On a transversal notion of ecology see F. Guattari, *The Three Ecologies*, London: Continuum, 2008; Code op. cit.; on an expansive concept of ecology see K. de Laplante. 'Toward a More Expansive Conception of Ecological Science' Biology and Philosophy, 19, 2004

12   Philippopoulos-Mihalopoluos 2013 op. cit.

13   Code op. cit. at 25

14   C. Douzinas and A. Gearey, *Critical Jurisprudence. The Political Philosophy of Justice*, Oxford: Hart Publishing, 2005; p. 50

15   P. Osborne. 'Modernity is a Qualitative, Not a Chronological, Category' in F. Barker, P. Hulme and M. Iverson (eds) *Postmodernism and the Rereading of Modernity*, Manchester: Manchester University Press, 1992, p. 23

16   Thus e.g. G. De Anna. 'Modernità e Immanenza: l'Azione Umana in Tommaso D'Aquino e Thomas Hobbes' in L. Parisoli (ed.) *Il Soggetto e la Sua Identità. Mente e Norma, Medioevo e Modernità*, Palermo: Officina di Studi Medievali, 2010; Osborne op. cit.

17   See e.g. B. de Sousa Santos. *Toward a New Common Sense. Law, Science and Politics in a Paradigmatic Transition*, London: Routledge, 1995; P. Grossi. *L'Europa del Diritto*, Bari: Laterza, 2001

18   de Sousa Santos op. cit.; see also Grossi 2001 op. cit.

19   I am thinking in particular of the State, and of the principle of sovereignty. The role of the State, and whether and to which extent it is waning, remains a debated question. See e.g. Hardt and Negri op. cit. and S. Sassen. *Territory Authority and Rights. From Medieval to Global* Assemblages, Princeton, NJ: Princeton University Press, 2006. On the question of sovereignty see e.g. M. Koskenniemi 'What Use for Sovereignty Today?' *Asian Journal of International Law*, 1:1, 2011. Arguments for the continuing centrality of sovereignty however are in e.g. J. Crawford. 'Sovereignty as Legal Value in J. Crawford and M. Koskenniemi (eds) *The Cambridge Companion to International Law*, Cambridge: Cambridge University Press, 2012

20   '[E]ver since its formation' would remark Foucault, modernity 'has found itself struggling with attitudes of "countermodernity"', M. Foucault. 'What is Enlightenment?' in M. Foucault. *Ethics: Subjectivity and Truth. Essential Works of Foucault 1954–1984, volume 1*, edited by P Rabinow, London: Penguin Books, 2000, p. 310

21   Thus e.g. Grossi 2001 op. cit.

22   The mythology of the origin being a typical modern obsession, see R. Esposito, *Living Thought. The Origins and Actuality of Italian Philosophy*, Stanford, CA: Stanford University Press, 2012. Thus also R. Orestano, *Introduzione allo Studio del Diritto Romano*, Milano: Giuffrè, 1987

23   Douzinas and Gearey op. cit. p. 50

24   See in general A. Gramsci, *Selections from the Prison Notebooks*, edited by Q. Hoare and G. Nowell-Smith, London: Lawrence and Wishart, 2007 and E. Laclau and C. Mouffe, *Hegemony and Socialist Strategy: Towards a Radical Democratic Politics*, 2nd edition, London: Verso, 2001; with particular reference to law see D. Litowitz, 'Gramsci, Hegemony and the Law', *Brigham Young University Law Review*, 2, 2000

25   Which finds a central referent in Plato's philosophy, which however has always

been co-existing with constantly marginalized alternative ways to interpret and construct the world. Recently alternative strands have gained increased attention, through the work of e.g. Næss (founder of deep ecology, recovering the legacy of Spinoza), Deleuze and Guattari, Hardt and Negri (also recovering the legacy of Spinoza) and Esposito (tracing the history an entire alternative strand of modern philosophy and political theory that he calls Italian Thought, and which includes philosophers such as Machiavelli, Vico, Bruno and, more recently, though along the same conceptual continuity, Croce, Gentile and then Gramsci, Negri, Agamben and Esposito himself), Esposito op. cit.

26   A. Grear 'The Vulnerable Living Order: Human Rights and the Environment in a Critical and Philosophical Perspective' *Journal of Human Rights and the Environment*, 2:1, 2011, 27

27   And here the inherently aporetic character of modernity is made visible, right at the interface between the mind has a double ontology: as both rational thought and material organ

28   Indeed, Descartes's goal was to 'purge the mind from the senses', and thus reduced knowledge to geometric reason and the certainty of mathematical demonstrations, L. Solidoro, 'Il Giudice e Il 'Fatto': Nuove Suggestioni del Pensiero Vichiano' Teoria e Storia del Diritto Privato, 6, 2013, 23, my translation. Hobbes would transfer this ideo-ontological commitment to political theory and law, see T. Hobbes, *Leviathan or the Matter, Forme, & Power of a Common-Wealth Ecclesiastical And Civill*. Printed by A. Crooke, at the Green Dragon in St. Paul's Churchyard, 1651

29   B. Latour, 'An Attempt at a "Compositionist Manifesto"' *New Literary History*, 41:3, 2010, 476

30   Here I combine the concepts of ideology and ontology to signify how they support one another: rather than ontology being an objective representation of the world, it is understood as being deeply entangled with a particular set of (normative) ideas about the world

31   P. Curry, 'The Third Road: Faërie in Hypermodernity', in G. Harvey (ed.) *The Handbook Of Contemporary Animism*, London: Routledge, 2013

32   See e.g. Tallacchini 1996 op. cit. and H. Jonas, 'The Scientific and Technological Revolutions', *Philosophy Today*, 15:2, 1972

33   Jonas op. cit., p. 76

34   Jonas op. cit., p. 80, highlighting the central contributions, in this respect, of Copernicus and Galileo

35   A. Hornborg, *The Power of the Machine. Global Inequalities of Economy, Technology, and Environment*, Altamira Press, 2001, p. 163

36   A. Grear, *Redirecting Human Rights. Facing the Challenge of Corporate Legal Humanity*, London: Palgrave-McMillan, 2009, p. 41

37   Horkheimer and Adorno suggested for example that 'domination' becomes 'objectified as law', M. Horkheimer and T. Adorno, *Dialectic of Enlightenment: Philosophical Fragments*, Stanford, CA: Stanford University Press, 2002, p. 37

38   G. Bascherini, 'Italian Theories. Spunti attorno all'esperienza giuridica a partire da un recente saggio di Roberto Esposito' Costituzionalismo, 1, 2013. Available online at: www.costituzionalismo.it/articoli/438/ (accessed 4 August 2014)

39 Laclau and Mouffe op. cit.; Litowitz op. cit.; Law is in this respect 'twice fragile', as Tigar and Levy underline, insofar as it is constantly subjected to a multiplicity of pressures (ranging from reformist to revolutionary), and this makes it all the more paramount to assert and re-assert hegemony, M. Tigar and M. Levy, *Law and the Rise of Capitalism*, Delhi: Aakar Books, 2005, p. 293

40 Litowitz op. cit. for example, discussing law in terms of hegemony, substitutes the Gramscian concept of class with the postmodern one of code; see also Tuori, which, drawing on the Foucauldian concept of 'episteme', calls this code the 'deep structure' of law, K. Tuori, *Critical Legal Positivism*, Farnham: Ashgate, 2002

41 Grear 2011 op. cit.

42 Law's contribution to these processes is well-known; see e.g. Holder op. cit.; Grear 2011 op. cit.

43 I draw this theoretical distinction from R. Thomas-Pellicer, *What is Kultur?: The Places of God in the Age of Re-embodiments*, PhD Thesis, Surrey: University of Surrey, 2012, which however developed the distinction within a broader sociological and social theoretical context

44 Grear 2009 op. cit. at 41

45 Grear 2009 op. cit. at 42

46 J. Nedelsky, *Law's Relations: A Relational Theory of Self, Autonomy, and Law*, Oxford: Oxford University Press, 2011, p. 163

47 I am referring in particular to the Franciscan philosophy of *dominium* and of the autonomous subject, see e.g. V. Mäkinen, *Property Rights in the Late Medieval Discussion on Franciscan Poverty*, Louvain: Peeters Publishers, 2002 and P. Grossi 'Usus Facti. La Nozione di Proprietà nella Inaugurazione dell'Età Nuova', Quaderni Fiorentini per la Storia del Pensioro Giuridico Moderno, 1, 1972, 287

48 There is no space to discuss all these manoeuvres. I will have to remain limited to a few brush strokes that highlight rather than analyse the disembodiment of the legal subject, and otherwise refer to existing literature, such as Grear 2009 op. cit., Nedelsky op. cit.; N Naffine, 'Who are Law's Persons? From Cheshire Cats to Responsible Subjects', *The Modern Law Review*, 66:3, 2003

49 The neuter being the perfect fit, to the extent that the abstract, disembodied subject is a neuter, as it leaves behind any form or particular material embodiment. However, it must be strongly emphasized that historically, the subject's material referent has been predominantly a rational male.

50 Indeed voluntarism is understood as quintessentially modern, M. Villey, *La Formation de la Pensée Juridique Moderne*, 2e triage, Paris: Quadrige/Puf, 2009

51 G. Teubner, 'Costituzionalismo Societario e Politica del Comune' in S. Chignola (ed.) *Il Diritto del Comune. Crisi della Sovranità, Proprietà e Nuovi Poteri Costituenti*, Verona: Ombre Corte, 2013, p. 48

52 E. Pashukanis, *Law and Marxism: A General Theory*, London: Pluto Press, 1989; C. Miéville *Between Equal Rights. A Marxist Theory of International Law*, London: Pluto Press, 2006

53 Grear speaks of quasi-disembodiment in order to signal the 'incomplete emptiness of the formally empty subject. It reveals the sleight of ideological hand invoked by the notion of abstraction', Grear 2009 op. cit. at 44

54 R. Esposito, *Le Persone e Le Cose*, Milano: Einaudi, 2014

55 Grear 2009 op. cit., Nedelsky op. cit.; Naffine op. cit.

56 Thus explicitly Hegel in his Philosophy of Right: 'property is the means by which I give my will an embodiment', G. Hegel, *Philosophy of Right*, trans. by S Dyde, Kitchener, ON: Batoche Books, 2001, para 46a

57 Thus e.g. D. Delaney, 'Making Nature/Marking Humans: Law as a Site of (Cultural) Production', *Annals of the Association of American Geographers*, 9, 2004, 487, p. 489

58 Weston and Bollier op. cit. at 128. This is not the place to explore the deeper genealogy of this epistemic imaginary, but it runs at least as deep as the dispute on poverty between the Papacy and the Franciscan minorite order that started in the thirteenth century; see e.g. B. Tierney, *The Idea of Natural Rights. Studies on Natural Rights, Natural Law and Church Law 1150–625*, Grand Rapids, MI: Eerdmans Publishing Company, 2001; P. Grossi, 'La Proprietà nel Sistema Privatistico della Seconda Scolastica', in P. Grossi (ed.) *La Seconda Scolastica nella Formazione del Pensiero Giuridico Moderno*, Incontro di Studi, Firenze 16–19 Ottobre 1972, Atti, Milano: Giuffré, 1972

59 E. Freyfogle *Justice and the Earth: Images for our Planetary Survival*, Champaign, IL: University of Illinois Press, 1996, esp. p. 49ff.; see also Weston and Bollier op. cit.; Cullinan op. cit.

60 This is the basic and fundamental mechanics of liberal law; see e.g. Pashukanis op. cit.; Miéville op. cit.

61 U. Mattei, *Beni Comuni. Un Manifesto*, Bari: Laterza, 2011, p. 59

62 Pieraccini op. cit. p. 17

63 In at least *some* of its forms, see e.g. Tallacchini 2000 op. cit.

64 D. Worster, *Nature's Economy: The Roots of Ecology*, 2nd edition, Cambridge: Cambridge University Press, 1994, p. 256

65 F. Golley, *A History of the Ecosystem Concept in Ecology: More than the Sum of its Parts*, New Haven, CT: Yale University Press, 1993; G. Mitman, *The State of Nature: Ecology, Community, and American Social Thought, 1900–1950*, Chicago: University of Chicago Press, 1992; V. De Lucia, 'Competing Narratives and Complex Genealogies. The Ecosystem Approach in International Environmental Law', *Journal of Environmental Law*, 27:2, 2015, 119, specifically in relation to how this complex genealogy manifests in international environmental law

66 A. Bell, 'Non Human Nature and the Ecosystem Approach. The Limits of Anthropocentrism in Great Lakes Management', *Alternatives Journal*, 20:3, 2004; De Lucia 2015 op. cit.

67 E. Darier (ed.), *Discourses of the Environment*, Oxford: Blackwell, 1999; Bell op. cit.

68 For a comparative discussion of two such legal philosophical approaches, see De Lucia 2013 op. cit.

69 Darier op. cit.

70 de Laplante 2004 op. cit.; Guattari op. cit.; Code op. cit.

71 Guattari op. cit. p. 29

72 Ibid. p. 28

73 Code op. cit. p. 19

74 Latour op. cit., p. 476

75   On the preference of the word 'world' over that of 'nature' see G. Cosi, 'Tutela del Mondo e Normatività Naturale' in G. Lombardi-Vallauri (ed.) *Il Meritevole di Tutela, Milano:* Giuffrè, 1990. In synthesis, world is preferable to nature since the latter is a modern category susceptible of logical analysis, while world is a word indicating an ecological complexity that resonates much better with the need to re-orient law in an ecological sense. This approach is quite close to the discussion of ecology in this article.

76   'Living' is intended here in its widest possible meaning, hence including also, say, rocks and other 'inert' bodies, under the understanding that each and every material body is an embodiment, a *somatic space* 'abundantly filled with meaning' (Thomas-Pellicer op. cit. p. 7, page number refers to chapter 4, single document on file with author), and possessing 'a lively materiality and agential significance of their own', Anna Grear, personal communication.

77   Code criticizes also merely 'contextual' understandings of knowledge, to the extent that, she suggests, in such accounts text and context can still be separated, Contextual accounts then fail to recognize the irreducible entanglement of knower and known in what she calls reciprocal relations of constitution, Code op. cit., p. 147, but also throughout, as this is a central thesis of Code's book, see e.g. the section called 'Situated knowing, Individualism, and Mastery', ibid. p. 133ff.

78   A. Philippopoulos-Mihalopoulos, 'Actors or Spectators? Vulnerability and Critical Environmental Law', Oñati Socio Legal Series, 3, 2013, 854; De Lucia 2015 op. cit.; A. Grear, 'Law's Entities: Complexity, Plasticity and Justice', *Jurisprudence,* 4, 2013, 76

79   Guattari op. cit., p. 25. In somewhat similar terms, Gramsci understood the individual person as being the result of multiple elements (ie the psychological, the social and the natural) and hegemonic pressures, making it impossible to speak of an individual as a coherent self-contained subject, see M. Filippini, 'Individuo e Individualità in Gramsci', *Critica Marxista,* 3–4, 2007, 35

80   Guattari op. cit. p. 25

81   Tallacchini 2000 op. cit., p. 1093

82   Smith 1997 as quoted in Tallacchini 2000 op. cit. p. 1092

83   M. Tallacchini 'Diritto, Complessità, Ecologia' in Lombardi-Vallauri op. cit.

84   Tallacchini 1996 op. cit., p. 326

85   See e.g. C. Stone 'Should Trees Have Standing? Toward Legal Rights for Natural Objects', *Southern California Law Review,* 45, 1972, 450, p. 456, and section 6.5 in this chapter

86   Tallacchini 2000 op. cit., p. 1094. We have seen how this is a major issue raised also by Earth Jurisprudence.

87   Tallacchini 2000 op. cit., p. 1094

88   Nedelsky op. cit.

89   K. Shrader-Frechette, 'Methodological Rules for Four Classes of Scientific Uncertainty' in J. Lemons (ed.) *Scientific Uncertainty and Environmental Problem Solving, Oxford:* Blackwell, 1996; Code op. cit.; K. deLaplante, 'Is Ecosystem Management a Postmodern Science?' in B. Beisner and K. Cuddington (eds) *Ecological Paradigms Lost: Routes of Theory Change,* Burlington, MA: Elsevier Academic Press, 2005

90 See e.g. S. Gutwirth and E. Naim-Gesbert, 'Science et Droit de l'Environnement: Réflexions pour le Cadre Conceptual du Pluralism de Vérités', Revue interdisciplinaire d'études juridiques, 34, 1995, 33; see also N. De Sadeleer, *Environmental Principles. From Political Slogans to Legal Rules*, Oxford: Oxford University Press, 2008

91 Tallacchini 2000, op. cit., p. 1095

92 Subjective biases may arise from personal biases; from social or cultural preferences; and/or from methodological choices that themselves carry an axiological dimension. Methodological operations such as extrapolation from one context to another 'are never neutral and univocal, but are always influenced by values and goals' Tallacchini ibid. at 1096. See also K. Shrader-Frechette, *Risk and Rationality. Philosophical Foundations for Populist Reforms*, Berkley, CA: University of California Press, 1991, and Shrader-Frechette 1996 op. cit., pp. 12–39

93 De Sadeleer op. cit., esp. p. 251

94 E. Fisher, 'Environmental Law as 'Hot' Law', *Journal of Environmental Law*, 25, 2013, 347, pp. 347–8

95 De Lucia 2015 op. cit. Some however still maintain that 'science has the answers', placing then a different kind of responsibility on law: that of implementing those answers, C. Voigt 'The Principle of Sustainable Development. Integration and Ecological Integrity' in Voigt op. cit., p. 153

96 Latour op. cit., p. 478

97 Tallacchini 2000 op. cit., p. 1095

98 Code op. cit., p. 5

99 iIbid., pp. 5–6

100 As Mattei has suggested, the birth of modernity is directly linked to the destruction of the commons, Mattei 2011 op. cit.

101 Arun Agarwal, quoted in Weston and Bollier op. cit., p. 146

102 Weston and Bollier op. cit., p. 173

103 Grossi consistently uses the plural form when referring to collective forms of appropriation, in order to emphasize the conceptual and ideological distance between those and modern private property, utterly singular, uniform and intolerant of diversity and plurality. See for example P. Grossi, 'La Proprietà e le Proprietà nell'Officina dello Storico', *Quaderni Fiorentini per la Storia del Pensiero Giuridico Moderno*, 17, 1988, 359

104 P. Grossi, 'Assolutismo Giuridico e Proprietà Collettive' Quaderni Fiorentini per la Storia del Pensioro *Giuridico Moderno*, 19, 1990, 505, p. 510

105 Thus Regnoli, an Italian borgeousie jurist, as reported in Grossi 1990 op. cit., p. 510

106 See e.g. Grossi 1990 op. cit. and P. Parajuli., 'Revisiting Gandhi and Zapata: Motion of Global Capital, Geographies of Difference and the Formation of Ecological Ethnicities', in M. Blaser, H. Feit and G. McRae (eds), *In the Way of Development. Indigenous Peoples, Life Projects and Globalization*, London: Zed Books, 2004

107 Weston and Bollier op. cit.; Mattei 2011 op. cit.; Hardt and Negri op. cit.

108 Civic and/or social commons are those urban commons such as public libraries, schools, parks, theatres, etc.

109 Digital commons are, for example, the Internet, various common resources such as Wikis and Open Source Software, etc.

110   Such as theatres, a prime example being the 'Teatro Valle' in Rome

111   Thus e.g. Marella 2012a op. cit.

112   See e.g. de Sousa Santos op. cit.

113   See e.g. V. De Lucia, 'Law as Insurgent Critique. The Perspective of the Commons in Italy', *Critical Legal Thinking*, 5 August 2013. Available online at: http://criticallegalthinking.com/2013/08/05/law-as-insurgent-critique-the-perspective-of-the-commons-in-italy/ (accessed 3 February 2015); U. Mattei 'Protecting the Commons: Water, Culture, and Nature: The Commons Movement in the Italian Struggle against Neoliberal Governance', *South Atlantic Quarterly*, 112:2, 2013, 366

114   See Weston and Bollier op. cit.; Hardt and Negri op. cit.; Mattei 2011 op. cit.

115   M. Marella, 'Introduzione', in M. Marella (ed.), *Oltre il Pubblico e il Privato. Per un Diritto dei Beni Comuni*, Verona: Ombre Corte, 2012, pp. 17–8

116   Thus Marella 2012 op. cit. p. 17. The 'problematic heterogeneity' is my own term, but tries to summarize Marella's position as regards the need for a taxonomy

117   Marella op. cit., p. 17

118   Marella 2012 op. cit., p. 17

119   Marella 2012 op. cit., p. 17

120   Marella 2012 op. cit., p. 20

121   Marella 2012 op. cit., p. 20

122   Weston and Bollier op. cit., p. 126

123   Weston and Bollier op. cit. , p. 126

124   As Godden observes, modernity's 'tendency to differentiate and classify [...] marks a discontinuity with earlier, more organic perceptions of the world', L. Godden, *Nature as Other: The Legal Ordering of the Natural World*, PhD Thesis, Faculty of Law, Griffith University, 2000, p. 30. On modernity as a classificatory order see also M. Horkheimer, *The Eclipse of Reason*, Oxford: Oxford University Press, 1947 and M. Foucault, *The Order of Things. An Archaeology of the Human Sciences*, London: Routledge, 2002

125   Weston and Bollier op. cit., p. 124 describe the commons as a matrix of perception and discourse, in the singular. I add the plural, as I think it better reflects that the commons are a plurality of practices, knowledges and forms of life, in this respect deeply resonating with Code's framework of epistemic location, resisting in this sense the modern 'epistemological monoculture', Code op. cit., p. 8

126   Weston and Bollier op. cit., p. 124

127   I refer in particular, and for reasons of familiarity, to the contemporary practices of the commons in Italy, see e.g. Mattei 2013 op. cit. These practices are however emerging everywhere in the Global North with similar political, legal and conceptual orientations throughout the world, see e.g. Weston and Bollier op. cit.

128   M. Foucault, *Society Must Be Defended. Lectures at the Collège de France 1975 – 76*, London: Penguin Books, 2003, p. 9. The extension to juridical forms is our own

129   Foucault 2003 op. cit., p. 7

130   Foucault 2003 op. cit., p. 7

131   Foucault 2003 op. cit., p. 7

132 See e.g. M. Marella, The Constituent Assembly of the Commons (CAC), OpenDemocracy 28 February 2014. Available online at: www.opendemocracy. net/can-europe-make-it/maria-rosaria-marella/constituent-assembly-of-commons-cac where she describes a special experiment ongoing in Italy of bottom-up constitutionalism, where legal scholarship, specialized legal expertise, militant activism, practices and contributions of commoning and active citizenship converge and contribute to the carving of new forms of legality for the commons. In fact, this constituent experiment 'has a two-level structure: it consists of: a) a travelling assembly which includes activists & jurists and gathers in different places in Italy; and b) a drafting commission (only jurists)'

133 Foucault 2003 op. cit.

134 P. Linebaugh, *The Magna Carta Manifesto. Liberties and Commons for All*, Berkley, CA: University of California Press, 2008

135 Linebaugh op. cit., p. 45 Linebaugh also insists upon the fact that commoning 'inheres in a particular field, upland, forest, marsh, coast', ibid., p. 45

136 Linebaugh distinguishes explicitly the two types of rights, Linebaugh. op. cit. p. 44

137 Linebaugh op. cit., p. 45. And while Locke linked private appropriation to the mixing of labour with natural things, in the context of the commons labour is not linked to private appropriation, but to access

138 O. Panigiani, *Vocabolario Etimologico della Lingua Italiana*, Milano: Albrighi and Segati, 1907, 'comune'. Available online at: www.etimo.it. Such obligations of participation intrinsic to the notion of commons are also interestingly present, reversed, in the term immunity, which originally means 'exempt from obligations'

139 Even the feudal and pre-modern concept of property was not a 'mere aggregate of economic privileges, but a responsible office', Tigar and Levy op. cit., p.183

140 See in this respect also P. Grossi, *Le Situazioni Reali nell'Esperienza Giuridica Medievale*, Padova: Cedam, 1968

141 Teatro Valle is the oldest operating theatre in Rome. In 2011, threatened with privatization, it was reclaimed as a commons through occupation by a group of artists that took over its management and offered, until their eviction in August 2014, a full artistic program to the people of Rome, and continuous political, legal, artistic and educational laboratories. For more details see e.g. S. Bailey and M. Marcucci, 'Legalizing the Occupation: The Teatro Valle as a Cultural Commons', *South Atlantic Quarterly*,112:2, 2013, 396

142 Linebaugh op. cit., p. 45

143 Linebaugh op. cit., p. 45. In this sense also P. Grossi, *L'Ordine Giuridico Medievale*, Bari: Laterza, 2006

144 Independence from the State is the last characteristic Linebaugh lists, Linebaugh op. cit.

145 Importantly, however, certain resources are immanently more susceptible to be claimed as commons, to the extent that they are necessarily implicated in the sustenance of life. There are no commons though without commoners, even if, as Rodotà underlines, the legal regimes proceeds from the resource, and not vice versa. Both Weston and Bollier op. cit. and Rodotà in this respect identify some necessary commons. Rodotà in particular considers commons

by necessity all those that are capable of providing utilities that are functional to the realization of the fundamental rights inscribed in the Constitution, see Rodotà Commission Bill, Delegated Legislation to Reform the Civil Code Articles Concerning Public Property, Act of the Senate of the Republic n. 2031, XVI Legislature

146 Mattei 2011 op. cit., p. 54
147 M. Marella 'La Parzialità dei Beni Comuni contro l'Universalismo del Bene Comune', 6 May 2014. Available onlin at: www.euronomade.info/?p=2282
148 Mattei speaks of a distinction between a language of 'being' and a 'language' of having. I rather speak of a language of 'doing and becoming' to avoid static representations of being susceptible of essentialization; the political intention of this differentiation however is similar, and it relates to fundamentally delineating the commons against the modern horizon of property (i.e. having) see Mattei 2011 op. cit.
149 Though not everyone agrees, see P. Maddalena, 'I Beni Comuni nel Diritto Romano: Qualche Valida Idea per gli Studiosi Odierni', Federalismi.it – Rivista Italiana di Diritto Pubblico Italiano, Comunitario e Comparato, 14, 2012. Available online at: www.federalismi.it/nv14/articolo-documento.cfm?artid=20444 that argues that formal title is very important
150 R. Rota, 'Sussidiarietà e Ambiente: la Centralità dell'Uomo'. Paper presented at the Conference 'Il governo del territorio e dell'ambiente per l'uomo', held on 22 June 2007 at Università of Rome 'Tor Vergata', p. 17
151 P. Barcellona, *Diritto senza Società: dal Disincanto all'Indifferenza*, Bari: Edizioni Dedalo, 2003, pp. 50–1
152 Linebaugh op. cit., p. 45; Grossi 2006 op. cit.
153 See on the commodity form of law Pashukanis op. cit.; Miéville op. cit.
154 Hardt and Negri op. cit., p. 121
155 Hardt and Negri op. cit., p. 122
156 Code op. cit., p. 7; Code also draws from Wittgenstein
157 Code op. cit., p. 7
158 Code op. cit., p. 8
159 Code op. cit., p. 50
160 See e.g. Hornborg op. cit. on space and time appropriation and the displacement of the ecological consequences of industrial technology
161 An increasingly crucial question that reflects a concept of law as a technique crucially imbricated with economics and capitalist calculations, see e.g. U Mattei and L Nader, *Plunder. When the Rule of Law is Illegal* (Blackwell 2008)
162 R. Esposito, *Pensiero Vivente. Storia e Attualità della Filosofia*, Milano, Italiana: Einaudi, 2012, p. 24
163 Thomas-Pellicer op. cit., p. 7 (page number refers to chapter 4, single document on file with author)
164 Guattari op. cit., p. 24
165 Esposito 2012 op. cit., p. 68
166 Philippopoulos-Mihalopoulos 2013 op. cit.
167 A. Grear, personal communication
168 The word aesthetics derives from the verb αἰσθάνομαι (*aisthanomai*) and from the noun αἴσθησις (*aisthesis*). They both refer to *awareness* and *perception*. How-

ever they both refer to *both* immediate sense perceptions (whether through hearing, sight, taste, scent or touch) as well as intellect or discernment. Aisthanomai is used in this double sense for example in the Bible, in Lk. 9:45 and 1 Tim. 2:3–4

169 S. Romano, *L'Ordinamento Giuridico*, 2nd edition, Firenze: Sansoni, 1946, p. 28, my translation. This is not the place for elaborating on Romano's theory. For an English account of his theory see F. Fontanelli 'Santi Romano and L'ordinamento giuridico: The Relevance of a Forgotten Masterpiece for Contemporary International, Transnational and Global Legal Relations', *Transnational Legal Theory*, 2:1, 2001, 67 and M. La Torre, *Law as Institution*, London and New York: Springer, 2010, especially pp. 98–115 where La Torre discusses Romano's theory and then compares it with MacCormick and Weinberger's (new) institutionalism

170 Romano op. cit., p.29, my translation

171 This translates Romano's 'ordinamento giuridico'. For problems relating to this translation see Fontanelli op. cit.

172 Fontanelli op. cit., p. 79

173 T. Schultz, *Transnational Legality: Stateless Law and International Arbitration*, Oxford: Oxford University Press, 2014, p. 61, citing Italian legal scholar Tarello. This position seems close to that taken by Weston and Bollier, to the extent that, while they recognize the normative capacity of the commons, they imagine it couched within and enabled by the structures of formal State law, Weston and Bollier op. cit.

174 Weston and Bollier however do not necessarily see antagonism between the commons and the State, Weston and Bollier op. cit.

175 Thus Romano op. cit., p. 8; Mattei 2011 op. cit.

176 There is a clear genealogical referent for this model, and namely the roman municipal republican model. There is no space to delve into this here, and I can only remand to some key literature, such as e.g. P. Catalano, *Populus Romanus Quirites*, Torino: Giappichelli, 1975 and G. Lobrano, *Res Publica Res Populi. La Legge e la Limitazione del, Potere*, Torino: Giappichelli, 1997

177 This is in many ways an inevitable consequence. Romano in fact derived legal pluralism from his institutional account of law

178 Philippopoulos-Mihalopoulos 2013 op. cit., p. 857

179 Philippopoulos-Mihalopoulos 2013 op. cit., p. 860

180 Modern subjectivity is instituted through a logic of presupposition, that is a presupposition of antecedence of the subject with respect to both the world of phenomena, and to itself as a body, Esposito 2012 op. cit.

181 Philippopoulos-Mihalopoulos 2013 op. cit.

182 M. Hasley, *Deleuze and Environmental Damage. Violence of the Text*, Farnham: Ashgate, 2006. This oscillation is pervasive, and has been at the root of heated debates in relation to biodiversity conservation for quite some time; see e.g. K. Kloor 'The Battle for the Soul of Conservation Science' *Issues in Science and Technology*, 2005. Available online at: http://issues.org/31-2/kloor/

183 Philippopoulos-Mihalopoulos 2013 op. cit., p. 860

184 Esposito 2012 op. cit.; but the productive potential of conflict was already recognized by Heraclitus, in what became a minority position within the

context of Greek philosophy, see e.g. M. Butz, 'Chaos Theory, Philosophically Old, Scientifically New', *Counseling & Values*, 39:2, 1995, 84

185   Bascherini op. cit.; Orestano op. cit.; Grossi 2006 op. cit.

186   This is probably particularly true for the movement of the commons in Italy; in this respect see especially Mattei 2011 op. cit., but in a more broader sense see also Hardt and Negri op. cit. The dimension of conflict is underexplored in Weston and Bollier op. cit., insofar as they see State as a necessary *enabler* of the commons (which they call, drawing on Ivan Illich, vernacular law), in the sense of providing the facilitating overarching legal structures within which vernacular law should be able to operate

187   See in general, Esposito 2012 op. cit., especially the part dedicated to Machiavelli

188   While the aim of analysis is that of 'tightening up' or 'narrowing' the meaning of a term (what Heidegger calls 'stunting the word'), etymology aims at '*opening up*' the word, in order to reveal the richness of its semantic field, M. King, 'Heidegger's Etymological Method: Discovering Being By Recovering The Richness Of The Word', *Philosophy Today*, 51:3, 2007, 278, p. 278

189   Panigiani op. cit., 'compire'. Available online at: http://www.etimo.it

190   Panigiani op. cit., 'complemento'. Available online at: http://www.etimo.it

191   This for example is the understanding of the principle of part-whole in roman law, according to Maddalena op. cit., see in particular p. 5

192   Vulnerability is an increasingly significant theoretical and juridical concept. There is no space to really address that here, but see e.g. M. Fineman and A. Grear (eds) *Vulnerability Reflections on a New Ethical Foundation for Law and Politics*, Farnham: Ashgate, 2014; or more specifically Grear 2009 op. cit. (on the role of vulnerability in constructing re-embodied legal subjects) and Philippopoulos-Mihalopoulos 2013 op. cit. (on the role of vulnerability in relation to a critical environmental law)

193   Pashukanis op. cit.

194   M. Vogliotti, 'Introduzione' in M Volgiotti (8ed), *Il Tramonto della Modernità Giuridica*, Torino: Giappichelli 2008) p. 6, my translation

195   Grear 2009 op. cit.

196   L. Fuller, 'Human Interaction and the Law', *American Journal of Jurisprudence*, 14, 1969, 1, p. 2

197   Fuller op. cit., p. 20

198   G. Postema, 'Implicit Law', Law and Philosophy, 13:3, 1994, 361; furthermore, this interactive dimension of implicit law is, according to Postema, 'the hallmark of Fuller's jurisprudence', p. 364.

199   Fuller op. cit., p. 3

200   Fuller op. cit., p. 3

201   Fuller op. cit., p. 11

202   Stone 1972, op. cit., p. 456

203   Grear 2011 op. cit.

204   This is for example the approach taken by Cormac Cullinan (personal annotation during the Conference 'Rule of Law for Nature', Oslo, 9–11 May 2012)

205   Tallacchini 1996 op. cit. speaks in this respect of a 'field effect', p. 337, my translation

206 Philippopoulos-Mihalopoulos 2013 op. cit., p. 866
207 Ibid., p. 866
208 Grear 2009 op. cit.
209 Particularly interesting is A. Timmer. 'A Quiet Revolution: Vulnerability in the European Court of Human Rights', in M Fineman and A Grear (eds), *Vulnerability Reflections on a New Ethical Foundation for Law and Politics*, Farnham: Ashgate 2013, discussing how the concept of vulnerability has found some traction in the jurisprudence of the European Court of Human Rights
210 And this is the focus of Earth Jurisprudence, namely the question of rights of nature, see e.g. Cullinan op. cit.
211 On this imbrication see de Sousa Santos op. cit.; see also, with particular relation to the construction of the liberal legal subject Grear 2009 op. cit.

Chapter 9

Graffiti artists and guerrilla gardeners
Challenging our understandings of Property Law

Sue Farran

Introduction

This paper is written from a legal perspective and in particular one based on the social and economic construction of property informed by western, liberal capitalist economies, nuclear families and legal systems in which the protection of property rights is central. The legal embodiment of property is informed by a set of norms accepted, willingly or not, by the society in which the law operates. A recurring theme in English property law is that the relationship between people and land is seen as manifesting itself in the form of rights, so that property is not so much about the soil or bricks and mortar, or about personal histories and identities, as about the rights which are appended to these: ownership, possession, use.[1] These rights in turn are shaped by legal frameworks that reflect certain implied or expressed agreed signifiers of importance, for example, the right of a possessor to exclude others; the right of an owner to freely alienate his or her property. Thus legal rights are supported by and in turn support various morally approved claims or forms of conduct. Central to these is the sanctity of private property and certainly since the late nineteenth/early twentieth century the importance of property, including land, as a marketable commodity.[2] As a consequence, one finds very early on laws that criminalise theft and damage to property, and civil laws, which provide remedies against trespass, nuisance, and interference with property. These private laws spill over into the public domain and shape the way in which the use and occupation of space and place is determined in public areas, streets, parks, shopping centres and so on. Recent decades however have seen new meta-norms emerging that may conflict with those currently reflected in property law. While there appears to be increasing restriction of conduct seen as anti-social or non-conformist,[3] and emerging threats to freedom of expression from concerns about terrorism,[4] there is also the emergence, or perhaps one should say 'revival', of non-individual rights or common rights, such as environmental rights, and incursions on the rights to various forms of private property through the emergence of concepts, such as the creative commons, open access, the right to have access to the countryside and the coast.[5] This move away from individualism to collective or communal rights echoes historical forms of communal land rights in England pre-dating the Enclosure Acts of the late-eighteenth and early-nineteenth century,[6] which were themselves triggers of protest,[7] and finds echoes in societies where land and people are more closely connected, both traditionally,[8] and today,[9] and where land is of much greater significance as a source of origin, marker of identity and mainstay of survival than a commodity for commercial exchange.[10]

It is against this shifting normative landscape that this paper considers two forms of engagement with property in largely – but not solely – urban environments: guerrilla gardening and graffiti 'art'.[11] Both guerrilla gardening and graffiti are spatial practices that not only 'draw attention to the complex processes at work in the social, cultural and political construction of urban space',[12] but also challenge the legal framing of that space and politics of control.[13] As Matless has stated 'Moral codes are revealed whenever their limits are transgressed'.[14] So an analysis of guerrilla gardening and graffiti reveals that the phenomena do two things: they expose the normative order underpinning the present embodiment of property within the law, and they challenge this by offering an alternative way of seeing and relating to property. These two forms of engagement: guerrilla gardening and graffiti are found in many countries.[15] In this paper I am interested in the ways in which they question the assumptions and socio-economic norms that underpin legal constructions of property as found in England and Wales. However, given the broad dispersal of western liberal legality, the analysis here may resonate in other jurisdictions in which there are similar debates about private/public ownership of space, communal enjoyment/economic capitalisation and identity of place/control of space.[16]

Understanding property

First, what do I mean by property, guerrilla gardening and graffiti and how are they linked? In the context of examining the impact of these last two on the legal embodiment of property I am using the term 'property' in a broad sense encompassing the interaction of law and property in criminal law and the private law of tort – as in damage to property and/or trespass; intellectual property – as in the rights of ownership of a creator; and property law – as in the law that relates to land and buildings. Guerrilla gardening has been defined as 'illicit gardening on someone else's land',[17] while graffiti has been defined variously as 'art', 'vandalism', and 'An element of the Hip Hop culture misinterpreted and misrepresented by the mainstream media, and most especially hated by affluent (usually white) businessmen who don't understand the roots or meaning of the writing on the walls'.[18] Both activities occur on property that is not legally owned by the graffiti artist of guerrilla gardener but which is usually accessible to the public – visually or physically or both.[19]

Neither guerrilla gardening nor graffiti are twenty-first century phenomena.[20] The guerrilla gardening movement in its present form first started in New York in the 1970s as part of the counter-culture movement of the time.[21] Today it might be seen as part of a much wider socio-cultural movement of individual and community engagement with gardening in diverse ways that reflect growing concern with the environment, with sustainable development, with the accountability of public authorities and with the retention or safeguarding of public space.[22]

Graffiti too has historical links and the word derives from the Greek *graphein* ('to write'),[23] or from the Italian *graffiare* (to scratch).[24] Modern graffiti is closely associated with socio-cultural, politico-art movements such as Situationist street art that emerged in Paris in the 1950s,[25] pop art in the 1960s,[26] which had origins in the United States, or hip-hop and rap in the 1970s, through to the present day,[27] which also had origins in New York,[28] but has become a global phenomenon. While ancient graffiti such as hieroglyphics are revered, modern attitudes to graffiti are more mixed. Protagonists have argued that graffiti can 'Enhance and alter its surroundings through a colourful explosion of geometric and serpentine shapes and colours'.[29] In some cases and in some places graffiti may be regarded as cultural heritage – for example in the Falls Road in Belfast, or indeed as mainstream culture – see for example the pieces of the artist Banksy,[30] whose work, it has been suggested in England, should be eligible to be listed (and therefore preserved) under the Planning (Listed Buildings and Conservation Areas) Act 1990.[31] Others have suggested that graffiti is no more than vandalism, anti-social behaviour and an attack on private property.[32] Negative stereotyping may be due to the failure to distinguish between different forms of graffiti, or because the line between what might be regarded as mere 'tagging'[33] and what is regarded as 'art' in the world of graffiti is contentious.[34] Reporting the 'theft' of a Banksy work from the wall of a shop, a spokesman for the Arts Council of England said: 'We believe all art, whether viewed in a gallery, museum or in the everyday urban environment, should be valued so it is a shame that a piece of street art that is well loved by the local community has been removed for auction. We understand that Haringey Council (the local authority of the location of the building where the artwork was located) is currently investigating how the removal occurred and we await the outcome of their investigation with interest.'[35] Not all share the view that graffiti is 'street art', not least because of the location of the expression. Banksy, the artist mentioned above, has himself displayed ambivalence about the art/vandalism debate, asking, 'Is graffiti art or vandalism? That word (art) has a lot of negative connotations and it alienates people, so no, I don't like to use the word "art" at all.'[36] An implication here is that what is valued as 'art' is itself very much a socio-economic construction with various creative works acquiring the status of 'art' through marketing, media and promotion by celebrities, elites and institutional players. Indeed the very term 'art' may be contrary to the political and social messaging behind graffiti, just as 'politically correct' terms such as 'greening the environment' or 'encouraging localism' might be distasteful to guerrilla gardeners. Diverging views, however, also suggest that there are no clear paradigms here.

Challenging the boundaries that distinguish the lawful from unlawful

The absence of clearly marked categorisation boundaries is reflected in the challenges posed by these activities for the law. Both manifest themselves in a wide variety of forms of engagement – in the case of graffiti this may range from 'tags' or 'throw-ups' and slogans to master-pieces of mural art ('pieces'); in the case of guerrilla gardening this may be random seed scattering, single plantings or large-scale gardening projects. This broad spectrum of activity is matched by porous boundaries between what is accepted and what is not accepted by society – and in turn law enforcement agencies. The locus of activity of guerrilla gardening and graffiti is likely to infringe the property rights of others, nevertheless, despite the potentially law-breaking nature of both activities, they may be adopted and 'legitimised' by the public or private sector: in the case of graffiti through commissioning, public approbation or the dedication of specific public and private surfaces for graffiti artists,[37] and in the case of guerrilla gardening by adoption of projects by local councils or other organisations. Consequently both are paradoxically potentially aesthetic, acceptable, practices and unacceptable, criminal acts. They are also, in their own right, and in many different ways, embodiments of diverse world views. For example, those interested in the artistic dimension of graffiti might ask 'is it art?', while those interested in the political dimension would perhaps focus on the use of art and/or graffiti as protest.[38] Similarly, while guerrilla gardening might be carried out as a form of protest,[39] carrying with it (negative) political overtones, it may also be viewed positively as a social and cultural phenomenon that brings people together either as individual members of the movement or as groups initiating community projects. Consequently the same normative ambivalence that surrounds graffiti attaches to guerrilla gardening: is it a constructive or destructive use of property, is it beautification or vandalism? Is it anti-social, because it is often secretive, or is it in fact socially constructive, bringing people with common concerns together? Diverse views present a challenge for the value systems that inform the law.

On the face of it guerrilla gardening and graffiti are illegal, in so far as they fall outside the law of property and within the criminal law or they attract liability in private law. Guerrilla gardening potentially involves the torts of trespass and nuisance as well as potential breaches of the criminal law. Nevertheless, prosecution of guerrilla gardeners in the UK seems very rare,[40] although guerrilla gardens may be destroyed either by officials or by vandals. The legal armoury of local authorities seeking to prosecute guerrilla gardeners who garden local authority land is also weak. While gardening on central reservations of roads, roundabouts or railway sidings may raise health and safety issues, ironically it is often the local authority that should be charged with health and safety offences for allowing land to become a hazard – for example, because of the rubbish fly-tipped on it, or the use to which it is put for drug-dealing and prostitution, or because noxious weeds are allowed to take hold.[41] Similarly, while the tort of trespass is available to landlords or owners of guerrilla gardened spaces, the quantum of damages is likely to be negligible, especially if the guerrilla gardeners have cleared the space of rubbish and enhanced it by planting. The tort of nuisance is also unlikely to be successfully pursued, especially where the gardeners may have removed a number of the pre-existing potential nuisances. It is also the case that urban development, the privatisation of formerly national assets, such as railways, and the re-demarcation of council and borough boundaries, leaves land orphaned. There are therefore, pockets of land that lie unclaimed or the ownership of which is unclear so there is no one to bring an action against a guerrilla gardener. There is also land that has been purchased for development, often by non-resident land investors or local councils, which is left undeveloped. So again, the matter arises as to who is to bring any legal claim, and it may be many years before the guerrilla gardening activity is challenged. Ironically, the improvements of the locality due to guerrilla gardening undertaken as a protest against private property or neglected public property may enhance the value of property in the area, putting at risk any undeveloped land and further excluding those who have no land. The 'adoption' of guerrilla projects by local councils or communities, or the official provision of alternative land for such projects may also terminate guerrilla status.

Unlike the case with guerrilla gardening where legal action seems rare, graffiti is much more likely to provoke a negative legal reaction. In English law, for example, graffiti is categorised as criminal damage,[42] and potentially attracts a punishment of imprisonment for adult offenders if the damage (usually assessed by the cost of removal of the graffiti) exceeds £5,000.[43] Minor graffiti offences are punishable by a penalty notice of £50 issued under the Anti-Social Behaviour Act 2003,[44] or a caution.[45] In other words while some graffiti is criminalised as being merely anti-social, other graffiti is criminalised as damage to property.

Criminalising graffiti is not without its problems. First, there is the matter of whether a specific piece of graffiti – or all graffiti (and the law makes no distinction) – is 'damage'. This term is not defined in the legislation but has been explained as being when the value or usefulness of a thing is impaired,[46] permanently or temporarily.[47] The unhelpfulness of this non-definition is evident in the decisions of the courts, where it has been held that whether there is or is not damage is a matter for the jury to decide so that graffiti *may* be damage, but equally it *may* not.[48] As Edwards points out 'the spatial contexts in which graffiti is situated needs to be considered if its social, aesthetic and legal significance is to be appreciated. They (the court decisions) also show the contested economic, social and political value of some publicly situated property'.[49] Second, there is the question of whether the graffiti artist had the necessary criminal intent to cause damage, or, alternatively, to create a work of 'art'. In other words is the subjective intention of the 'artist' the defining factor, or is the status of the work determined objectively by reference to an external set of socio-legal norms? As pointed out by the Banksy quote above, for many the term 'art' may have negative connotations and be rejected by the artist even though others may refer to the work as 'art'. This might particularly be a factor when considering graffiti 'pieces'. If the test is objective then the graffiti artist's own subjective belief in the value or purpose of his/her works is irrelevant. If however it is subjective the outcome could be different. In other words, if Banksy does not think of his work as art or refer to it as such because he dislikes that term, is it nevertheless art? Third, could the aesthetic enhancement of an ugly hoarding or boarded-up warehouse afford a 'lawful excuse' for an otherwise potentially criminal act, thereby providing a defence to the charge?[50] Under the relevant criminal law a 'lawful excuse' rests on the belief of consent.[51] On the face of it this suggests there is some communication between the owner of the property 'damaged' and the graffiti artist, but such is the nature of graffiti that this is unlikely except where work is commissioned or surfaces provided for graffiti. But if the artist is an eminent graffiti artist then he/she may presume consent – the graffiti will after all enhance the value of the building, or, if the surface is an eyesore owned by the local council the consent of the public might be presumed either because the site is already a graffiti space or because the 'piece' had turned an eyesore into a work of art. Similarly there is scope to interpret the provisions of the law so as to accommodate subsequent public approbation of this damage to publicly owned buildings found in the phrase: 'would have so consented if he or they had known of the damage...' For example, Bristol City Council conducted a public poll subsequent to the completion of a 'piece' to ask if the public wanted the graffiti to be removed or to remain. Overwhelmingly they wanted it to remain.[52]

The legal reaction to graffiti seems to be premised on two factors: the sanctity of private property, and the belief that graffiti is aberrant to normal social behaviour.[53] Criminalising or legally punishing an act places it outside the accepted normative order. In fact so great is the stereotyping of graffiti as vandalism and/or anti-social behaviour that in England public authorities may order graffiti to be cleaned off private property even if the owner of that property wishes to retain the graffiti.[54] Yet some boroughs and councils set aside graffiti areas known as 'free walls' or adopt a no-prosecution policy in respect of certain areas, which become 'graffiti tolerance zones',[55] or identify graffiti as local or national heritage.[56]

Similarly, attitudes towards prosecution appear to vary, and those empowered to issue penalty notices – Police Community Support Officers, British Transport Police and those persons accredited under a community safety scheme – may chose not to exercise those powers, or may appear to distinguish between different forms of graffiti, demonstrating a greater willingness to prosecute taggers,[57] or cases of graffiti on property where the victims of such graffiti can be clearly identified, for example, private house owners or commuters where trains have to be taken out of service,[58] than sites where there is no evidence of active ownership. There is also inconsistency in sentencing: for example if the graffiti is not 'threatening or abusive' sentence may be more lenient,[59] and similarly if there is no evidence that the graffiti 'caused harassment, alarm or distress to anyone'.[60]

It is not just the application of the criminal law that is problematic. There are also dilemmas in intellectual property law. In principle the creator of a work of graffiti has intellectual property rights to that work, provided graffiti is regarded as a 'work' (Copyright, Designs, and Patents Act 1988, section 4(1) applies in UK)[61] and provided the author/creator can be identified and is not anonymous – which may occur because of the potential legal penalties authorship might attract,[62] or, even if 'signed', the identity of the artist may be unknown – as in the case of Banksy. Although these may only be moral rights,[63] in so far as economic rights might be frustrated by the illegal nature of the creation,[64] these are still rights recognised in law, and the law itself makes no reference to whether the act of creativity has to be legally done. Moreover, with art extending to encompass installations, public art and performance art, it has been suggested – at least in the Banksy's case – that the removal of the mural and its sale by auction could be a continuum of the original artwork.[65] Thus any subsequent interference could lead to claims of intellectual property infringement. The implication of a continuing intellectual property right was manifest in 2014 when Banksy gave his permission for one of his works – originally attached to the building but subsequently removed for safe-keeping, to be sold by a youth centre in Bristol to raise funds.[66] Alternatively, any act aimed at the physical property that damages the graffiti could be viewed as a breach of the copyright right of integrity of the work. Potentially therefore there could be a clash of ownership interests between the owner of a wall, for example, and the graffiti artist whose work adorns that wall.

Whether the graffiti artist or indeed the guerrilla gardener has any intellectual property rights in their creation is an interesting question because there is nothing in the law to prevent them claiming copyright.[67] If someone then takes a photo of the work and commercially exploits it, the only objection to the original creator successfully defending his or her copyrights may be moral objections on the grounds that the activity was illegal in the first place. It might also be argued that the premises underlying copyright law – to incentivise and reward the creative author of the work – are misplaced in the context of either guerrilla gardening or graffiti and that existing laws, here as in other areas, have limited relevance, so that to deny copyright is in line with normative underpinnings of graffiti or guerrilla gardening which challenge the property status quo.[68]

Denying property rights recognised in law, however, means that graffiti designs can be freely copied for wallpaper, advertising and other purposes,[69] while guerrilla projects can be purloined or capitalised on by local authorities seeking to establish 'green' or 'community-friendly' credentials without acknowledging the contribution of the guerrilla gardeners. The sites of graffiti and guerrilla gardening are therefore ones of contestation, not just in a physical sense but in a moral sense as well: their very existence challenges the foundations on which understandings about property are constructed.

At the same time, however, graffiti and guerrilla gardening are creative acts that make a statement through the medium of interaction with physical property – land and other surfaces – although the maker of the statement may be disenfranchised from property rights or perceive themselves to be so: graffiti artists may be overrepresented by ethnic minorities,[70] or socially marginalised groups;[71] guerrilla gardeners may themselves be landless, for example, because they live in urban high-rise buildings or dense urban housing. So property as physical space is important to them while at the same time their activities are often on or at locations that are seen to be impersonal or as 'negative space', a term used by graffiti artists interviewed by Halsey and Young in their studies in Australia.[72] The interaction with this space is physical and moral: both may be using this 'space' for articulations of political and aesthetic views. For example, it has been stated that guerrilla gardening is 'a battle for resources, a battle against scarcity of land, environmental abuse and wasted opportunities. It is also a fight for freedom of expression and for community cohesion'.[73] Gardens may be planted to provide food,[74] to serve as memorials,[75] to draw attention – by comparison – to the ugliness of buildings, to provide an educative or leisure space or to enrich the biodiversity of the urban landscape. Similarly, graffiti may be the marking of territory by gangs,[76] the co-operative project of a 'crew'[77] or a form of political, racist or ethnic protest, social comment, ways of stating and asserting identity or reaction against imposed values and systems. For example, it has been suggested that

> Modern street art is the product of a generation tired of growing up with a relentless barrage of logos and images being thrown at their head everyday, and much of it is an attempt to pick up these visual rocks and throw them back.[78]

A landscape of changing values

The re-embodiment of this space is therefore dependent upon the existing framing of property and property rights in order to react against it. Indeed it could be argued that graffiti and guerrilla gardening are identified – in part at least – by their anti-authoritarian/ anti-establishment image. Both may also be vehicles for expressions of self as well as community, and modes for self-empowerment drawing on characteristics of daring, creativity, stealth, visibility and leaving a legacy (a platform of fame) – however short-lived.[79] At the same time, because graffiti and guerrilla gardening are manifested as markers in respect of place and space, surfaces are presented as sites, subjects and objects of communication, as statements and questions – so that property is simply the vehicle for the message. These messages delivered through gardening and graffiti may not go so far as to stake claims of ownership but they certainly make claims of space and place: 'I was here, this is my mark'. In other words these might be viewed as personal claims establishing a relationship with surfaces outside the legal framework of property interests, whereby individuals or groups identify through participation their 'place' in the landscape. Even when the graffiti artist or the gardener has moved on, they remain, if only temporarily, located at the site of their endeavours.

Buildings, parks, abandoned building sites cease to be property but become, through intervention, socio-political comment and the embodiment of a new counter-narrative, reminding us of the social capital in our cities, towns and villages through the interaction of people and landscape. The impact of this is conveyed in a passage from 'Re-creation: Phenomenology and Guerrilla Gardening'. The author writes of her reaction in seeing a clump of daffodils among urban rubble:

> Their instant beauty compelled me to attend to them in and of themselves, which I could only do by noting their contrast to the vacant lot and, thus, rethinking their interrelation with me in order to consider our cooperatively created meaning. Their appearance

challenged the everyday framework of meaning by which I approach the world.[80]

It is this challenge to the 'everyday framework of meaning' that may lie behind some of the outrage expressed about graffiti and guerrilla gardening. The need to 'clean up' graffiti, for example, may be symptomatic of a wider concern. As stated by the judge in the graffiti case of *R v Pease*: 'defacement of the urban landscape added to anxiety faced by society as a whole'.[81]

Behind this statement lies the implication that the normative order is under siege and that if this is not resisted the whole fabric of society is in danger of disintegration. In other words, maintaining the physical integrity of the urban landscape according to a certain set of normative values is essential for maintaining the integrity of something much more ephemeral. This normative order however reflects particularised interests in respect of property, which in turn are premised on assumptions about what socio-economic norms are shared or universally accepted. Graffiti and guerrilla gardening physically invade the boundaries of urban space, pay scant attention to the rules of property and demonstrate an opposing or different urban view. They are in short, invasive urban outlaws.

The 'wrongness' of this alternative approach would be much clearer if the lines between the acceptable and the not-acceptable were more succinct, but as indicated above, they are not,[82] which makes these adversaries of the established order difficult to confront or contain, physically, morally or legally.[83]

This porousness of boundaries has further significance. While these activities could be viewed as crimes and property incursions in so far as they engage with the horizontal and vertical surfaces of the urban environment and change them, they are also illustrative of an ongoing clash of views about the purpose and function of physical space and the rights and interests that are associated with these. As suggested above one alternative view might be that graffiti and guerrilla gardening make positive, aesthetic environmental statements, which challenge the environmental crimes of those who not only have property rights but also the power to improve the urban space for the greater benefit of those who live there. While it would be naïve to suggest that all graffiti artists or guerrilla gardeners subscribe to a shared world view, it might be argued that their engagement with physical space is a visual narrative that should open our eyes to possible alternative ways of seeing property and our relationship with it.

Time for a new lawscape?

It might therefore be asked whether it is not time for a new lawscape,[84] which more readily embraces property-related activities, which are currently marginalised or omitted.[85] For example, Edwards points out that in the criminal law 'Graffiti's ambiguous cultural status initially appears *not* to be reflected in its legal status',[86] and Webster has pointed out that if a Banksy was to be protected from removal by being listed, then the whole or part of the building would have to be listed as there is no scope within the relevant legislation to list the 'piece'.[87] Similarly, Davies has pointed out that intellectual property law, notably copyright, has limited application to the work of graffiti artists,[88] although Rychlicki suggests otherwise.[89] A similar dilemma affects guerrilla gardens. While garden designs might be copyrighted, the labour of making a garden can only confer notional Lockean rights.[90] Where guerrilla gardeners might succeed is not in claiming any form of property rights, but in asserting social or democratic rights in the form of community rights of engagement in local issues, buildings and land. Ironically given its guerrilla status, there have been advances in conferring more autonomy on communities in England and Wales in recent years through the Localism Act 2011, the Commons Act 2006 and, more recently, the Community Right to Reclaim Land.[91] Implicitly agreeing with some of the philosophy of guerrilla gardeners the government minister at the time stated:

> It's completely unacceptable that people have to walk past derelict land and buildings every day, in the knowledge that there's almost no prospect they will be brought back into use, and there's absolutely nothing they can do about it. For years, communities who have attempted to improve their local area by developing disused public land and buildings have found themselves bouncing off the walls of bureaucratic indifference – their attempts to do something positive for their community thwarted by a system that has proved totally ineffective.[92]

While this is unlikely to diminish guerrilla gardening or indeed graffiti, it does perhaps signal a reframing of the relationship between people in communities and the physical space of their location. Just as guerrilla gardening and graffiti may reflect a process of democratisation of property the previous UK government's advocacy of 'the Big Society',[93] and legislation in England and Wales in the form of the Localism Act 2011, supports this articulation of opinion from local people about the urban environment and the property in it.[94] The Act also requires local authorities to list assets and land 'of community value' (sections 87, 88 and 105), suggesting that a more democratic process of categorising land is envisaged. One aspect of this 'Big Society' agenda that resonates with the messages of graffiti artists and guerrilla gardeners is that the voices of communities should be heard and to listen to these those in authority have to reach and engage with people from diverse backgrounds and holding diverse values. Whether this advocacy of greater community engagement to inform the politics and policies of urban environments will materialise remains to be seen.[95] The political rhetoric however suggests a rethinking, possibly not consciously, of the relationship between people and things.

English property law has conventionally focused on people, and the triangulated relationship in which the rights and obligations vis-à-vis the owner and the non-owner in respect of the thing are far more important than the thing itself. Hohfeld, for example, explains that property rights are not so much about the property but about the relationship between people in respect of that property.[96] While property rights cannot be totally extricated from social, communal and historic factors,[97] conventionally in western contexts property is isolated from place,[98] and the law has provided a supportive context for the principle that 'property and land are alienable and tradable'.[99] Property law has in consequence become abstracted from the materiality of the subject to which it applies.[100]

This 'dephysicalisation' and 'placelessness'[101] of property has however been challenged in recent decades by the emergence of environmental law, concerns about biodiversity and the sustainable management of natural resources, and cultural heritage associated with the landscape, in other words the relationship between people and places, and the importance of the materiality and locus of 'things', such as land and buildings. Graffiti artists and guerrilla gardeners illustrate this challenge through their physical interaction with land and bring to the fore the association of people with place. I would suggest therefore, that guerrilla gardening and graffiti offer a re-embodiment of property, which in turn demands a re-embodiment of property law: a new lawscape.

'Lawscape' is a term found in anthropological studies of indigenous peoples' relationship with land. Social anthropologists refer to 'legalscape' or 'lawscape' to describe the interaction between people, land and the law. In many traditional societies, the location of people in the landscape is a complex and multifaceted relationship in which the land and its resources are not seen as commodities attracting marketplace claims, but as spiritual and temporal indicators of identity. As these societies and their economies change new legal forms and language are adopted, modified and adapted to reflect the changing relationship of people with land. In developing societies this process involves drawing on a range of legal resources including traditional land customs and laws and new forms of legal regulation to re-embody the relationship of people and place. The examples used in this paper suggest that this process need not be limited to developing societies but that a reverse or mirror re-embodiment is taking place through the activities of graffiti artists, guerrilla gardeners and indeed others who are engaging unconventionally with the landscape. If as Hayek suggests, the law, such as that found in England and Wales and the United States of America i.e. the common law, is the favoured instrument of capitalism because of its predilection for freedom of contract and its protection of private property,[102] then perhaps these examples of anti-capitalist expression in respect of property are merely two among many on a spectrum of property-related activities that challenge the legal systems which underpin this focus on economic interests.[103]

Conclusion

In this paper I have focused on two creative 'outlaw' urban activities that affect the vertical and horizontal surfaces of urban environments and challenge the law that determines ownership and control of these spaces. I have also drawn attention to the problems of propertising certain forms of creativity using existing legal frameworks, such as intellectual property law or planning law. While the criminal law has always been solidly grounded in the protection of property, especially private property, its enforcement in the context of creative acts may be problematic when ambivalent approaches are apparent and the applicable normative orders are inconsistent. The rhetoric behind the sanctity of private property is non-plussed when confronted by creativity in the form of gardening or graffiti 'pieces'. The certainty of the law wavers if distinctions are made between different forms of activity that may range from mindless acts of vandalism to careful and considered visual statements – in the case of graffiti – or from small innocuous single plantings to the take-over of whole roundabouts or gardening sites by guerrilla gardeners. The integrity of the established embodiment of property law is further complicated by the wider context of contemporary social conscience evident in concerns about: the aesthetic improvement of the built environment in inner city areas; improvement of brown field sites and/or their retention as natural habitats for urban flora and fauna; encouraging the engagement of local communities – particularly groups tending to disengage with society; and the green agendas of local government. This is not to suggest that all human endeavours should be brought within the law but to highlight that what might be learnt from the activities of creative outlaws is that the law has its limits and it is the spaces at the edge of the law that may be the most interesting. It is from the perspectives on the margins that the possibilities of re-embodying the law in a new form might be considered.

Notes

1   An overview of different theories of 'property' can be found in N. Glackin, 'Back to bundles: deflating property rights, again' (2014) 20(1) *Legal Theory* 1–24, although Glackin offers his own take on the matter.

2   See D. Cowan, L. Fox O'Mahony and N. Cobb, *Great Debates in Property Law*, London: Palgrave Macmillan, 2012, pp. 47–48; T. Nicholas, 'Businessmen and land ownership in the late nineteenth century' (1999) 1 *Economic History Review*, 27–44.

3 Evidenced by the use of anti-social behaviour orders under the Crime and Disorder Act 1998, the criminalisation of trespass under Part V Criminal Justice and Public Order Act 1994, and measures available under the Anti-Social Behaviour Crime and Policing Act 2014. See A. Crawford, 'Criminalizing Sociability through Anti-social Behaviour Legislation: Dispersal Powers, Young People and the Police' (2009) 9(1) *Youth Justice,* 5–26; A. Cornford, 'Criminalising Anti-Social Behaviour' (2012) 6(1) *Criminal Law and Philosophy,* 1–19.

4 See for example Article 19, *The Impact of UK Anti-Terror Laws on Freedom of Expression,* submission to the International Court of Justice (ICJ) Panel of Eminent Jurists on Terrorism, Counter-Terrorism and Human Rights, London, April 2006; I. Cram, *Terror and the War on Dissent: Freedom of Expression in the Age of Al-Qaeda,* New York: Springer, 2014.

5 See for example J. Mitchell 'What public presence? Access, commons and property rights' (2008) 17(3) *Social and Legal Studies,* 351–67.

6 See for example S. Bilsborough, 'A Hidden History: Communal Land Ownership in Britain' (1995) 16(3/4) *Ecos Magazine.* Available online at: www.caledonia.org.uk/land/bilsborough.htm (accessed 11/03/2015).

7 See for example C. J. Griffin, *Protest, Politics and Work in Rural England 1700–1850,* London: Palgrave Macmillan, 2014, although not all agree see J. Stevenson, *Popular disturbances in England 1700–1832,* 2nd edition, Abingdon: Routledge, 1992.

8 For example, B. Okot writing about Uganda, states, 'In most of Uganda, land equates to history, heritage, identity, belonging, rights and relationships' 'Uganda: Breaking the links between land and the people' International Institute for Environment and Development, 11 March 2012. Available online at: www.iied.org/uganda-breaking-links-between-land-people (accessed 11/03/2015).

9 See J. Perera (ed.), *Land and Cultural Survival: the Communal Land Rights of Indigenous Peoples in Asia,* Manila, Asian Development Bank, 2009; B. Feiring (ed.), 'Indigenous peoples' rights to lands, territories and resources', Synthesis Paper, International Land Coalition 2013, Rome. Available www.land coalition.org (accessed 11/03/2015).

10 This is widespread in indigenous communities see, for example, K. Göcke, 'Protection and Realization of Indigenous People's Land Rights at the National and International Level' (2013) 5 (1) *Goettingen Journal of International Law,* 87–154.

11 Whether graffiti is art or vandalism or somewhere between the two is too big a field to engage in here. See M. Halsey and A. Young, 'The meanings of graffiti and municipal administration' (2002) 35(2) *Australian & New Zealand Journal of Criminology,* 165–86.

12 C. McAuliffe and K. Iveson, 'Art and Crime (and Other Things Besides ...): Conceptualising Graffiti in the City' (2011) 5 (3) *Geography Compass,* 128–143, 129.

13 See K. Iveson, 'Cities within the City: Do-It-Yourself Urbanism and the Right to the City' (2013) 37(3) *International Journal of Urban and Regional Research,* 941–56.

14 D. Matless, 'Moral geographies' in Johnston, R. J., Gregory, D., Pratt, G. and Watts, M. (eds) *The Dictionary of Human Geography,* Oxford: Blackwell, 2000, pp. 522–3.

15   See for example N. Ganz, *Graffiti World: Street Art from Five Continents*, London: Thames and Hudson, 2009 and R. Reynolds, *On Guerrilla Gardening: a Handbook for Gardening without Boundaries*, London: Bloomsbury, 2008.

16   See for example, D. Tracey, *Guerrilla Gardening: A Manualfesto*, Gabriola Island: New Society Publishers, 2013; A. Waclawek, *Graffiti and Street Art (World of Art)*, London: Thames and Hudson, 2011 and R. Schacter and J. Fekner, *The World Atlas of Street Art and Graffiti*, London: Aurum Press, 2013.

17   Reynolds above, 5.

18   This is the definition taken from the *Urban Dictionary*. Compare the *Oxford Dictionary*, which defines the term as 'writing or drawings scribbled, scratched, or sprayed illicitly on a wall or other surface in a public place'.

19   Railway sidings or the sides of bridges, for example, may be difficult to access physically but be visible to passengers on trains.

20   See for example M. Hardman, 'Understanding guerrilla gardening: an exploration of illegal cultivation in the UK', Working Paper Series No. 1, School of Property, Construction and Planning, Birmingham City University, 2011.

21   M. Fraser, 'How guerrilla gardening took root' *BBC News* (Scotland), 15 March 2010. Available online at: news.bbc.co.uk (accessed 21/09/2015).

22   S. Farran, 'Earth Under the Nails: The Extraordinary Return to the Land' in Hopkins, N. (ed.) *Modern Studies in Property Law* 2013, Oxford: Hart, 173–91; O. Zanetti, *Guerrilla Gardening: Geographers and Gardeners, Actors and Networks: Reconsidering Urban Public Space*, MA/MSc dissertation 2007. Available guerrillagardening.org (accessed 22/09/2015).

23   S. Phillips *The Dictionary of Art*. London: MacMillan, 1996.

24   M. Spocter, 'This is My Space: Graffiti in Claremont, Cape Town' (2004) 15(3) *Urban Forum* 292–304 at 294; DeNotto, M. 'Street Art and Graffiti: Resources for online study' (2014) 75(4) *College and Research Libraries News*, 208–211.

25   See M. Wark, *The Beach Beneath the Street: The Everyday Life and Glorious Times of the Situationist International*, London, Verso, 2011. At the time an anarchic and loosely structured movement, the philosophy of the Situationists still inspires. See M. Battersby, 'The artist vandalising advertising with poetry', *The Independent*, 3 February 2012, writing of the (now) respected Scottish artist Robert Montgomery.

26   See, E. Elmaleh, 'American Pop Art and political engagement in the 1960s' (2003) 22(3) *European Journal of American Culture*, 181.

27   See for example hip-hop protests over the shooting of black teenager Michael Brown in the USA in August 2014.

28   But see McAuliffe and Iveson op. cit. for the larger picture.

29   Nigel Blunt, UKGraffiti.com quoted on: *The Site*. Available online at: www.thesite.org/homelawandmoney/law/yourrights/graffitiandthelaw (accessed 20/09/2015).

30   ICLR, 'Banksy under the hammer: graffiti and the law', Incorporated Council of Law Reporting, 6 March 2013. Available online at: www.iclr.co.uk/banksy-under-the-hammer-graffiti-and-the-law/ (accessed 24/09/2015).

31   J. Webster, 'Should the Work of Banksy be Listed?' (2011) *Journal of Planning and Environmental Law*, 374.

32   See for example, 'Vandalism, graffiti and environmental nuisance on public transport' in *Literature Review of Graffitti*, London, Her Majesty's Stationery

Office, 2004. Some simply reflect the confusion, see A. Akbar and P. Vallely, 'Graffiti: Street art – or crime?', *The Independent*, 16 July 2008. Available at: independent.co.uk (accessed 24/09/2015), or can see two sides to the issue, see for example 'Speaking with: Cameron McAuliffe on graffiti, art and crime', *The Conversation*, March 24, 2015. Available at: the conversations.com accessed 24/09/2015; J. Stewart, 'Graffiti Vandalism? Street Art and the City: some considerations' Unesco Observatory, Faculty of Architecture, Building and Planning, the University of Melbourne Refereed E-Journal, 86. Available at: education.unimelb.edu.au (accessed 24/09/2015).

33   The artistic presentation of 'signatures'.

34   See for example N. Sanchez, 'Graffitti: Art through Vandalism: A look at the Urban Art Movement through the Scope of Aesthetics and Illegality' Communication Over the Internet, (undated) University of Florida. Available online at: iml.jou.ufl.edu (accessed 24/09/2015) and N. Patel 'Graffiti in Austin: Crime of Art?', April 2010. Available online at: whatisart320. wordpress.com (accessed 24/09/2015).

35   R. Luscombe, 'Banksy mural mystery deepens as it heads for sale in Miami Arts Council in the UK, where Banksy mural went missing, petitions for return of artwork as auction moves forward', *The Guardian*, 20 February 2013. Available online at: www.theguardian.com/artanddesign/2013/feb/20/bansky-mural-slave-labour-miami-auction/print (accessed 10/08/2015).

36   Quoted by McAuliffe and Iveson op. cit. 130.

37   See for example, most recently the Dundee Graffiti Jam, L. Piper, 'Top aerosol artists head to city of Dundee Graffiti Jam', STV Dundee 17 June 2015, Dundee.stv.tv (accessed 23/09/2015), but compare G. Ogston writing about the same city earlier in the year 'West End of Dundee hit by 'Paris' graffiti tag', *The Courier*, 3 January 2015.

38   M. Dragićevi-Šešić, 'The Street as Political space: walking as Protest, Graffiti and the Student Carnivalizaton of Belgrade', (2001) 17(1) *New Theatre Quarterly* 74–86; E. Pindado, 'Graffiti artists turn Mexico City's walls into a collective cry of protest' *Fusion*, 13 March 2015. Available online at: fusion.net (accessed 23/09/2015); S. Jacobs 'The graffiti in Greece shows just how angry its citizens really are', *Business Insider UK*, 24 June 2015.

39   G. McKay, *Radical Gardening: Politics, Idealism and Rebellion in the Garden*, London: Francis Lincoln Limited, 2011.

40   See for example B. Daniel '"Guerilla gardeners" in Morpeth risk prosecution to carry out clean up', 29 September 2014, *ChronicleLive*. Available online at: chroniclelive.co.uk (accessed 23/09/2015), in which the local council said they would not be prosecuting gardeners who tidied up two neglected historic council buildings. See also E. Meskhi, 'Guerilla Gardening Good for Unused Land' (2010–2011) *The Innovation Diaries*. Available online at: the innovation-diaries.com (accessed 23/09/2015). Similarly a search of reported cases on the legal database WestLaw revealed very few cases in which prosecutions were brought.

41   Various weeds are classified as either noxious or invasive and if not eradicated by land owners can lead to prosecution under the Wildlife and Countryside Act 1981, or the issue of notices under the Weeds Act 1959 or the Town and Country Planning Act 1990/Town and Country Planning Act (Scotland) 1972.

42   Although not a specific offence in English law, graffiti falls under section 1 of the Criminal Damage Act 1971, that states 'A person who without lawful excuse destroys or damages any property belonging to another intending to destroy or damage any such property or being reckless as to whether any such property would be destroyed or damaged shall be guilty of an offence'. Damage that is evidently less than £5,000 is tried as a summary offence (subject to a fixed fine) in the Magistrates Court (Magistrates' Courts Act 1980). Where the damage is less than £5,000 the maximum sentence is three months imprisonment or a fine of £2,500 for adult offenders.

43   For those aged 12–17 years the maximum custodial penalty is a detention and training order of up to 24 months.

44   In England and Wales, section 48 of the Anti-Social Behaviour Act 2003 gives local authorities the power to serve graffiti removal notices on certain bodies responsible for the surface where graffiti has appeared. These bodies include the owners of street furniture (bus shelters, street signs, phone boxes, etc.). The notice gives a minimum of 28 days for the removal of the graffiti, if after that time it has not been removed the local authority can remove it and can recover its costs. It is also an offence to sell spray paint to under-16s. If a shopkeeper fails to prove they took reasonable steps to determine the age of the person, they can be fined up to £2,500. As graffiti may be done with other materials other than spray paint this is of rather limited value as a deterrent. In England and Wales the Clean Neighbourhoods and Environment Act 2005 enlarges the powers of local authorities and the police.

45   Guidelines issued by the UK Government suggest a caution for the minor offence or writing graffiti on a bus shelter. Available online at: www.gov.uk/caution-warning-penalty (accessed 23/09/2015).

46   *R v Fiak* [2005] EWCA Crim 2381.

47   *Morphitis v Salmon* [1990] Crim LR 48.

48   See the cases cited by I. Edwards, 'Banksy's Graffiti: A Not-so-Simple Case of Criminal Damage '(2009) 73 *Journal of Criminal Law*, 345–361, 350.

49   Edwards ibid., at 351.

50   Edwards ibid.

51   s. 5(2)(a) of the Criminal Damage Act 1971, a lawful excuse may be raised if: 'at the time of the act or acts alleged to constitute the offence he believed that the person or persons whom he believed to be entitled to consent to the destruction of or damage to the property in question had so consented, or would have so consented to it if he or they had known of the destruction or damage and its circumstances...'. The belief does not have to be justified.

52   See Webster op. cit. 374.

53   Reinforced by central and local government initiatives to encourage others to 'grass' on graffiti artists, through offering rewards for naming 'tags' or offering a 'hotline' for anonymous tip-offs.

54   Under the Town and Country Planning Act 1990 a local authority can serve notice on the property owner if, in the authority's opinion, the graffiti it is detrimental to the amenity of an area (s215). Local authorities in London have similar powers, Webster op. cit. 375.

55   In 2011 in Bristol for example the city hosted the UK's largest ever permanent street art project 'See No Evil' and see reference above to the Dundee graffiti

jam, and 'The rise of the Camden Street Art Scene' 2 March 2015, *Londonist*. Available online at: londonist.com/2015/03/the-rise-of-the-camden-street-art-scene.php (accessed 11/03/2015).

56 'Banksy's graffiti is part of the nation's heritage and deserves protection by law, claim academics', *Daily Mail*, 22 August 2011. Available online at: www.dailymail.co.uk/news/article-2028715/Banksys-graffiti-art-protected-law-say-Bristol-academics.html#ixzz2f4ic2PZC (accessed 11/03/2015), after one of his works was scrubbed off a wall in Bristol. Speaking in 2014 the Mayor of Bristol said 'He (referring to Banksy) is part of what gives Bristol its artistic, creative and subversive spirit which makes us such a sparky place' Flanagan op. cit.

57 See for example, the comments of the prosecutor in a 2013 case: 'We are not talking about witty imaginative images such as those I expect you are familiar with by Banksy, … What you are dealing with here is simple damage.' Quoted in *The Express*, April 13, 2013. Available online at: www.express.co.uk/news/uk/391442/Accounts-boss-is-250-000-rail-graffiti-vandal (accessed 11/03/2015).

58 See the judicial comment in *R v D* [2009] 2 Cr App. R (S) 74 para 24: 'This type of offending sickens members of the public who have their travelling lives blighted by this sort of criminal damage and vandalism by graffiti on trains on a massive scale. Sentences of course must be deterrent sentences to send a message out to those who may be tempted to do this for their own gratification'.

59 *R v Moore* [2011] EWCA Crim 1100.

60 *R v Thomas James Dolan and Thomas Albert Whittaker* [2007] EWCA 2791. Compare however *R v Brzezinski* [2012] EWCA Crim 108, where a sentence of 18 months' imprisonment for criminal damage caused by graffiti was upheld by the Court of Appeal, Criminal Division.

61 Now amended in parts by the Intellectual Property Act 2014.

62 Anonymous work may still be subject to copyright for a statutory period of time – see, section 12(2) Copyright, Designs and Patents Act 1988 in UK.

63 Section 78 Copyright, Designs and Patents Act 1988.

64 See J. Davies, 'Art Crimes? Theoretical Perspectives on Copyright Protection for Illegally Created Graffiti Art', (2012) 65(1) *Maine Law Review*, 28.

65 See ICLR op. cit.

66 Banksy gave permission to the Broad Plain and Riverside Youth Project to take ownership of the piece 'Mobile Lovers', P. Flanagan, 'Banksy's boost for Bristol youth club', *The Telegraph*, 15 April 2014. Available at: telegraph.co.uk (accessed 14/11/2014). The mayor of the city is reported to have said 'I hope it will be respected and protected as we would want for *any other legitimate work of street art*' (my italics). See also I. Johnston, 'Banksy breaks cover to join debate over 'Mobile Lovers' artwork', *The Independent*, 8 May 2014. Available online at: independent.co.uk (accessed 14/11/2014).

67 See for example, G. Friedman, 'Can Graffiti be Copyrighted', September 21, 2014, *The Atlantic* theatlantic.com accessed 23/09/2015; E. Bonadio, 'Graffiti copyright battles pitch artists against advertisers', 9 August 2014, *The Conversation*. Available online at: the conversation.com (accessed 23/09/2015).

68   In a forthcoming publication however M. Iljadica suggests that graffiti artists develop their own rules and practices to protect their work: *Copyright Beyond Law: Regulating Creativity in the Graffiti Subculture* Basingstoke: Hart, 2016.

69   See T. Rychlicki, 'Legal questions about illegal art' (2008) 3 (6) *Journal of Intellectual Property Law & Practice*, 393–401.

70   See J. Ferrell, *Crimes of Style: Urban Graffiti and the Politics of Criminality*, New York: Garland Publishing New York, 1993.

71   A. Nayak, 'Race, affect, and emotion: young people, racism, and graffiti in the post-colonial English suburbs', (2010) 42 *Environment and Planning*, 2370–92.

72   Halsey and Young op. cit. 286.

73   Reynolds op. cit.

74   See for example, M. Hardman and P. Larkham, *Informal Urban Agriculture: the Secret Lives of Guerrilla Gardeners*, London: Springer 2014.

75   Reynolds, for example, refers to the planting of pansies to commemorate homophobic attacks.

76   This seems to be a particular concern in the United States, see for example, S. Phillips, *Wallbangin': Graffiti and Gangs in L.A.*, Chicago, IL: University of Chicago Press, 1999.

77   See for example the 'Gospel Graffiti Crew ggcrew.org in California, 'Girls on top' crew' girlsontopcrew.blogspot.co.uk in the UK and the DDS Crew who specialise in trains: 'Underbelly: graffiti artists the DDS Crew paint London tube-trains- video', *The Guardian*, 9 July 2012. Available at: the guardian.com (accessed 24/09/2015).

78   Attributed to Banksy in 'The writing on the wall', *The Guardian* (Australia), 24 March 2006. Available online at: www.theguardian.com/artanddesign/ 2006/mar/24/art.australia (accessed 11/03/2015).

79   The self-focus of graffiti artists and guerrilla gardeners attracts criticism as much as the consequences of their actions on property, see for example, M. Hardman and P. Larkham, *Informal Urban Agriculture: the Secret Lives of Guerrilla Gardeners*, London: Springer, 2014, 185–92.

80   M. Walton, 'Re-creation: Phenomenology and Guerrilla Gardening' in R. Clingerman and M. Dixon (eds), *Placing Nature on the Borders of Religion, Philosophy and Ethics*, Farnham: Ashgate, 2011, 67 at p. 73.

81   *R v Matthew Pease and Others* [2008] EWCA Crim 2515.

82   J. Confino, '"Fake street art sucks" Perrier replaces Williamsburg's Nelson Mandela mural' *The Guardian* 26 September 2014; T. Pearson 'Is street art still "street"?' Vandalog – A Viral Art and Street Art Blog, 19 March 2011. Available online at: https://blog.vandalog.com/2011/03/is-street-art-still-street/ (accessed 11/03/2015).

83   For example, research in Australia in Graffiti 'hot spots' in Sydney indicated that cleaning did not reduce the incidence of graffiti but might cause its re-location: B. Haworth, E. Bruce and K. Iveson 'Spatio-temporal analysis of graffiti occurrence in an inner-city urban environment' (2013) 38 *Applied Geography* 53–63.

84   Lawscape is used in a number of different context to convey the spatial dimension of the reach and scope of the law. See for example, essays in I. Braverman, N. Blomley and D. Delaney (eds), *The Expanding Spaces of Law*, Stanford, CA: Stanford University Press 2014.

85 N. Lacey, C. Wells and O. Quick, *Reconstructing Criminal Law: Text and Materials*, 3rd ed. London: Butterworths, 2003, 314.

86 Edwards op. cit. 361.

87 Webster op. cit. 376. Even then, using the existing law would require some creative thinking.

88 J. Davies, 'Art Crimes? Theoretical Perspectives on Copyright Protection for Illegally Created Graffiti Art', (2012) 65(1) *Maine Law Review* 27–55.

89 T. Rychlicki, 'Legal questions about illegal art', (2008) 3 (6) *Journal of Intellectual Property Law & Practice*, 393–401.

90 Locke's theory was that a person deserved to keep the fruits of their labours, hence the development of private property rights. J. Locke, *Two Treatises of Government* 1689 (available through the Gutenburg online project www.gutenberg.org/files/7370/7370-h/7370-h.htm).

91 Announced by the Housing Minister on 2 February 2011 and facilitated by the Localism Act 2011, although a report in February 2015 suggests that few communities have exercised this right: L. Ryan, 'Community Rights: where's the evidence'. Available online at: opendemocracy.net (accessed 24/09/2015).

92 Above.

93 'Building the Big Society' was a concept of the coalition government of the UK aimed at putting 'more power and opportunity into people's hands' Cabinet Office UK Government. Available online at: www.gov.uk/governemnt/uploads/.../building-big-society_0.pdf (accessed 24/09/2015).

94 The Act includes provision for neighbourhood plans that could enable local people to propose a Street Art Policy for a particular Neighbourhood Area. Such policy could set out further provisions for retention or removal of graffiti and a mechanism for public consultation.

95 It has been suggested for example that guerrilla gardening could be valuably harnessed for urban regeneration: C. Palamar, 'From the Ground Up: Why Urban Ecological Restoration Needs Environmental Justice' (2010) 5(3) *Nature and Culture* 277–98 (and arguably graffiti could fulfil a similar role) and that greening the urban environment may in fact reduce crime: F. Kuo and W. Sullivan, 'Environment And Crime In The Inner City. Does Vegetation Reduce Crime?' (2001) 33 *Environment and Behavior*, 343–67.

96 W. N. Hohfeld, *Fundamental Legal Conceptions As Applied In Judicial Reasoning And Other Legal Essays* (Walter W. Cook ed., 1923).

97 Hepburn, for example, states that 'property is a social dynamic; mutable, mercurial and value laden' S. Hepburn, *Principles of Property Law*, Sydney and London: Cavendish, 1998, 2.

98 N. Graham, *Lawscape: Property; Environment and Law*, London: Routledge, 2011, 7.

99 Graham above: 10.

100 Graham above: 124.

101 Terms used by A. Layard, 'Book Review: Lawscape: Property; Environment and Law by Nicole Graham', (2011) *Journal of Environmental Law*, 160.

102 F.A. Hayek, *The Constitution of Liberty*, 1960 London: Routledge and Kegan Paul, 156.

103 See further S. Farran above.

Chapter 10

Autonomous legal persons and interconnected ecosystems

An 'Ecological' Self towards the age of re-embodiment

Alessandro Pelizzon[1] *and Gabrielle O'Shannessy*[2]

The dephysicalisation of law

Notwithstanding Oliver Wendell Holmes famous assertion that 'the common law is not a brooding omnipresence in the sky',[3] Western jurisprudence still envisions law as an abstract entity located somewhere outside the individual or any collective of individuals.[4] Legal positivism – in particular the identification of law as a relatively independent closed logical system[5] – still dominates the landscape of legal education as well as of legal practice, powerfully enshrined within Hart's five main tenets of positivism[6] and Kelsen's 'pure theory of law'.[7] But the idea of law as an abstract reality somehow extrinsic to – if not bereft of – human agency is not only the province of the positive 'science' of jurisprudence as envisioned by Austin.[8] Equally, natural law theories also locate the law somewhere 'out there', whether in a moral universe preceding human consciousness or in an inevitably uncertain and yet purportedly objectively discoverable 'state of nature'.[9] Western legal discourse thus appears engaged in a self-referential dialogue with its own abstract premises, removed from the physicality of the world in which that dialogue is inscribed.[10]

The modern construction of law as a disembodied set of rules with only a tenuous relationship with their own content is never more visible than in the abstract and rarefied conceptualization of property. Yet, this is exactly the terrain where the intrinsic unsustainability of such dominant jurisprudential view of law becomes ever more apparent. To Hohfeld, for example, ownership is about a 'legal relationship' with things, rather than a possession of things.[11] Even further, ownership is about a relationship between individuals with only an indirect reference to the very things that are the object of said ownership. Such is the argument articulated by Nicole Graham in her book *Lawscape*. While exploring 'the relationship between the abstractness of property law and the physical materiality of place',[12] the author argues that although the law, and in particular land law, ought to acknowledge a direct relationship to the environment through property regimes, such relationship – which, the author suggests, exists whether acknowledged or not – is instead obscured by the anthropocentric and self-referential nature of the current legal discourse. Owners have often no direct connection to the things they own other than the abstract idea of 'ownership' enshrined in the property regimes abstractly accepted by the existing legal discourse.

Foundational to the disconnection between the abstract conceptualizations of law and the objects to which these abstractions refer is an ontological distinction directly inherited from Cartesian philosophy. Pottage and Mundy indeed suggest that '[t]he distinction between persons and things may be a keystone of the semantic architecture of Western law.'[13] This distinction follows directly from the ontological differentiation between nature and culture, what Fitzpatrick calls the 'mythology of modernity'.[14] Graham writes that '[t]he paradigm of nature/culture operates via the dichotomous logic of anthropocentrism ... [which] divides the world into two categories: human beings and "the rest" ... [placing] humans at an imaginary centre of that world'.[15] Furthermore, 'people are not human in the sense of being a physically determined species – rather they are human in the sense of being a culturally determined and distinguished species from all other uncultured species.'[16] Therefore, cultured humans are thought to be separate from, and superior to, the category of nature. Empowered by Cartesian ontological possibilities, the separation of humanity from the world – or, more precisely, the many worlds – that humanity inhabits has caused the illusion of unfettered human mastery over the world itself.

This mastery becomes re-conceptualized in legal discourse through the distinction of the category of 'subjects' of rights – that is, persons, in their natural sense as well as legal constructs – and the category of 'objects' of such rights – that is, the rest of the world, the somehow inert collection of 'things'. The world, which largely belongs to the category of 'nature', thus becomes progressively more rarefied in legal discourse. Graham writes that 'law constructs itself as a metaphysical discourse that simultaneously constitutes and is constituted by the absence of the physical'.[17] Furthermore, 'this "dephysicalisation" of the world began by dividing the people-place relationship into the active agents of the property relation, "people", and the passive objects of the property relation, "things".'[18]

The physical and environmental effects of this abstract distinction are, however, hardly disputable. The 'dephysicalised' world of 'things' that are the 'objects' of rights in the current legal discourse constantly reasserts its presence as essential to the very existence of such a discourse, which would ultimately cease to have meaning if that 'dephysicalisation' were to be complete. Furthermore, property regimes authorized by the existing legal discourse, as Graham ultimately suggests, shape and alter the landscapes that we inhabit, transforming them into veritable 'lawscapes', a neologism coined to represent the inextricable connection between 'the law' and the world that it purports to regulate. In other words, the ontological possibilities awarded by Cartesian philosophy – and further embraced by Lockean individual liberalism – did not and do not transform the world by themselves. Rather, industrial societies articulate the horizon of ontological possibilities granted by determinist mechanistic anthropocentrism within legal and normative structures and mechanisms that allow natural landscapes to be physically changed, plied and shaped by the collective effort of humans, groups and institutions.

Swimme and Tucker remind us that

> Modern industrial humans… did not seek to commune with nature … They sought to transform the world. …The paradox of unintended consequences is now becoming evident… We have crossed over into an Earth whose very atmosphere and biosphere are being shaped by human decisions… We live on a… planet now, where not biology but symbolic consciousness is the determining factor for evolution. Cultural selection has overwhelmed natural selection. That is, the survival of species and entire ecosystems now depends primarily on human activities. We are [thus] faced with a challenge no previous human has ever contemplated: How are we to make decisions that will benefit an entire planet for the next several millennia?[19]

It is hard to deny that, by shaping the world we inhabit (and with it our very bodies as a consequence), law inheres in that world and cannot be separated from it. Even more importantly, to acknowledge that law lives within the world that it purports to regulate, that it is embodied in the very 'objects' of its abstract discourse, appears imperative in a world whose fate is determined by human activities and by the way in which such activities are regulated – that is, by law. The need of reconstituting the conceptualization of law is thus an ecological and economic imperative, one which is essential, it may be argued, to human survival.

Re-embodiment and earth jurisprudence

The challenge to the view of law as an abstract entity somehow separate from the lives and experiences of the people that are bound by it may very well have begun when Oliver Wendell Holmes asserted, in the 19th century, that 'the actual life of the law has not been logic: it has been experience'.[20] American legal realism famously began to break with a modern liberal jurisprudential tradition that discursively located law 'out there', whether in nature, in impersonal rules, or in a reality other than the lived experience of the individuals engaged with it. Davies aptly comments that '[t]he real nature of the law cannot be explained by formal deductive logic, only by its empirical and historical existence and the social ends to which it is directed.'[21] Furthermore, the author asserts, 'law is embodied, as we are, and denying this is only a way of making life and law easier by pretending that rules are an absolute justification.'[22]

Contemporary post-modern jurisprudence would certainly agree with this sentiment.[23] Rosen elegantly writes that

> law is so inextricably intertwined in culture that, for all its specialized capabilities, it may, indeed, best be seen not simply as a mechanism for attending to disputes or enforcing decisions, not solely as articulated rules or as evidence of differential power, and not even as the reification of personal values or superordinate beliefs, but as a framework for ordered relationships, an orderliness that is itself dependent on its attachment to all other realms of its adherents' lives.[24]

Douzinas and Geary indicate that '[p]ostmodern legality defies the [traditional] jurisprudential image.'[25] Be it Foucault's interest in the interaction of law and power,[26] Derrida's deconstructive philosophy,[27] or the *standpoint* epistemology advocated by some feminist scholars,[28] the voices challenging the 'metaphysics of modernity'[29] underpinning traditional Western jurisprudence are joined in a chorus that can hardly be ignored.

Furthermore, critiques to the abstractness and disembodiment of law do not come only from within the Western legal tradition, but also from without. Irene Watson voices an Aboriginal Australian perspective of law, according to which law is not simply a force that externally influences individual behavior but rather is an intrinsic part of the individual as well as residing in what the Western dichotomy identified above defines as 'nature':

> Our voices were once heard in light of the law. The law transcends all things, guiding us in the tradition of living a good life, that is, a life that is sustainable and one which enables our grand-children yet to be born to also experience a good life on earth. The law is who we are, we are also the law. We carry it in our lives. The law is everywhere, we breathe it, we eat it, we sing it, we live it.[30]

More recently, a new branch of legal theory, still in its infancy but nonetheless rooted in centuries of environmental legal and ethical thought,[31] is advancing a radical challenge to the traditional idea of law as an internally coherent and radically exclusionary corpus of rules, as a 'body' that exists outside of and without the physical bodies it professes to apply to. When Christopher Stone asked his provocative question in 1972, 'should trees have standing?'[32] his argument was innovative in that it proposed to remove the environment from the category of 'things', of 'objects' of someone's property rights. Stone's argument was further explored, two decades later, by Thomas Berry.[33] As a deep ecologist, Berry recognized an intrinsic value to nature itself, regardless of any human utility ascribed to it. Furthermore, Berry argued, if we are all part of an interconnected and interdependent system of beings and phenomena, then the wellbeing of each member of the system is connected to and dependent from the wellbeing of the system as a whole, and thus the safeguard of the wellbeing of the system is more important than that of each individual member of the system – such as humans. In other words, a *biocentric* or *ecocentric* perspective is preferable, in long-term ecological terms, to an *anthropocentric* one. The result of this paradigmatic shift in perspective led Berry to advocate a new 'Earth Jurisprudence'.

Cormac Cullinan, who coined the term 'Wild Law' to further advance Berry's suggestion,[34] describes Earth Jurisprudence as

> a philosophy of law and human governance that is based on the idea that humans are only one part of a wider community of beings and that the welfare of each member of the community is dependent on the welfare of the Earth as a whole. From this perspective, human societies will only be viable and flourish if they regulate themselves as part of this wider Earth community and do so in a way that is consistent with the fundamental principles that govern how the Universe functions (the 'Great Jurisprudence').[35]

Furthermore, Cullinan suggests, the *wilderness* is not the place of randomness and chaos that is often portrayed to be. On the contrary – and echoing Lovelock's theory[36] – it is a place of ultimate balance and order, a place of perfect homeostasis that ultimately transcends any human attempt to minutely regulate it.[37] Consequently, our human regulatory mechanisms – our 'earth governance' systems – should learn from the 'Great Jurisprudence' of the wilderness and should be designed in accordance with our understanding of it if long term human sustainability is to be achieved through legal means.[38] It is important to note that while Cullinan's comment aims to challenge the negative discursive connotations of the term 'wilderness', the idea of nature as a place of intrinsic balance is also not uncontroversial.

Some practical implications of Earth Jurisprudence may be perceived, at a first glance, as almost doctrinal and somehow inscribed within the classical debate between legal positivism and natural law theories. Cullinan's writings have certainly prompted legislation that, in a number of jurisdictions, has effectively removed 'nature' from the category of legal 'objects' and that has instead cast it as a legal 'subject' *of* and *with* rights, as in the cases of the many local ordnances drafted by the Community Environmental Legal Defense Fund in the United States, of the 2008

Ecuadorian Constitution and of the 2010 Bolivian Law of the Rights of Mother Earth.[39] However, the philosophy of Earth Jurisprudence – or Ecological Jurisprudence as some authors prefer to refer to it[40] – has much more profound implications than its practical implementation initially reveals.

Some commentators have analyzed Cullinan's suggestion of a dialogue between the 'Great Jurisprudence' and human governance as a novel form of natural law theory, particularly focusing on the implications of the 'laws of nature' for human jurisprudence. Burdon, for example, writes that '[Cullinan] regard[s] reference to the "laws of nature" as too general and too overwhelmingly broad to have relevance in Earth Jurisprudence'. Instead, '[Burdon] contend[s] that the Great Law should be defined with reference to "first principles" uncovered in the scientific discipline of ecology'.[41] However, other than shifting the goal posts of what the 'laws of nature' are, the comment fails to acknowledge the constructivist implications of Cullinan's and Berry's suggestions. By stating that '[t]he Great Law has been described as ontologically prior [to] and the measure of Human Law',[42] Burdon appears to simply restate in modern terms the classical Thomist distinction between the *lex naturalis* – that portion of the divine *lex aeterna* participated in by all rational creatures – and the *lex humana* – or positive law.[43] The 'law' is still, in this view, somewhere 'out there'.

Other authors, however, have instead began to explore the 'realm of creative uncertainty' offered by Earth Jurisprudence.[44] Schillmoller and Pelizzon acknowledge that '[a]t this stage in its development, both the practice and theory of this emergent jurisprudence occupy indeterminate terrain... one already inscribed by humanist precepts of what "rights" and "nature" might consist of',[45] and since 'nature and rights are contested concepts with negotiable meanings... [t]he consequences of such contestation for an earth justice system predicated on rights for nature is a problematic tension between the requirement of a concept of nature upon which to ground action, and an awareness of the impossibility of settling upon a definitive version of what nature "is"'.[46] While recognizing the intrinsic uncertainty of a new jurisprudence that stresses the interdependence between humans and the environment, between 'nature' and 'culture', the authors also acknowledge the creative possibilities contained in the emerging discourse of Earth Jurisprudence, with its inevitable gestures toward a re-interrogation of the ontological foundations of legal theory.

Far from perpetuating either the *separation* thesis or its antithesis,[47] the 'ought' and 'is' of the law are reconstituted within a broader cognitive discourse. Earth Jurisprudence invites scholars to engage with cognitive definitions of 'nature' by constantly negotiating and re-negotiating its meaning. The traditional dichotomy of 'nature' and 'culture' thus vanishes within a myriad of discursive practices and interactions that are nonetheless interconnected by engagement with the ultimate 'other', represented by 'nature' re-inscribed as a subject and as an active participant within the human political and legal world. Earth Jurisprudence is, therefore, intrinsically 'post-postmodernist', by which we mean here that while acknowledging that meaning is inherently contextual and contingent, 'nature' – or 'reality' – is not excluded from the system of cognition as inevitably unknowable, but rather it is acknowledged as a fundamental and irreducible element of discourse, one that forces us to constantly face deep metaphysical questions in our engagement with the law, questions that highlight the ultimate inseparability of 'the law' from the world within which it is embodied. Furthermore, the dialogues offered by Earth Jurisprudence offer a terrain for a cross-cultural encompassing theory of law, one that is informed by the multiplicity of *nomospheres* that paint the canvas of the diverse legal traditions of the world.[48]

A reconsidered self

In opening a creative space through which to question the metaphysical foundations of traditional jurisprudence by engaging with ontologically inseparable ecological considerations (that is to say, inseparable from the discursive domain of law), the discourse of Earth Jurisprudence becomes immediately relevant to transcendence of the 'dephysicalisation of law' discussed at the beginning of the chapter. Graham reminds us that 'the ongoing practice of property law depends entirely on a very particular, instrumentalist value of the biosphere' and '[t]his view or philosophy of the natural world or "nature" legitimizes current modes of... production, distribution and consumption of those "things"'[49] that constitute the natural world. Equally, it is a very specific and culturally inscribed idea of 'self' that underpins, facilitates and justifies these modes of production, distribution and consumption. It is this underlying concept of self within a liberalist ideology that empowers – if not creates – the current anthropocentric viewpoint and precludes a thicker ontological understanding of humans as interconnected beings within the world as a whole.

The liberalist idea of an autonomous and independent self forms indeed the basis for current notions of legal personhood, whether the abstract and disembodied legal 'person' is represented by a corporate entity or by a sovereign state. Davies and Naffine further affirm that they 'not only regard the legal person as a form of legal fiction, but [they] also [think] that there is a tacit view of the person that underwrites the official fiction, and that it too is a legal invention, one that is of course strongly influenced by social and cultural assumptions about the nature of a human being'.[50] The authors poignantly point to the idea of the self as a very specific cultural construct, one thoroughly clothed in liberalist assumptions, not only in light of the legal construct of an abstract 'self' such as an abstract entity but also in relation to the natural persons that form the fabric of any liberalist society.

In order to trace the origin of the notion of the autonomous self, it is necessary to trace the development of liberal ideology and positivism underpinned by the scientific beliefs and philosophy of the 17th Century from which that notion emerged. Prior to the 17th Century Roszak describes a naturalist view in which the self was seen to be inherently encompassed within nature.[51] It is with Locke's articulation of 'natural' rights – and in particular of 'natural' property rights – that the notion of an autonomous and independent self coalesced in its present form. Graham writes that 'Locke's justification of private property asserted natural rights and the wisdom of the market against divine rights and the wisdom of the kings. His appeal to nature worked as an escape route via which he could maintain the idea of origin as the guarantor of law's legitimacy while displacing the conventional connection of origin to force.'[52] It is particularly through the theory of John Locke that the notions of the autonomous person, property, and rights discourses have become inextricably intertwined.[53] Locke's concept of personhood and its relationship both to the body and to property has been the subject of much further study. His theory of 'ownership of the body and [its] labour'[54] was further developed in the writings of Hegel and Radin, who explored the theme of 'property for personhood' and the extension of 'self' through the outward accumulation of property.[55]

Locke's theory formed the basis for an idea of self as an autonomous, independent entity ultimately separated from the rest of the world. Through Lockean philosophy, Western law has thus thoroughly embraced the principles of Cartesian dualism, and the acceptance of mind as separate to matter, and, therefore, to the body, has thus become axiomatic within normative discourse. Ricketts argues that the abstraction that occurred from shifting from a moral discursive basis of law to a 'rights' basis of law allowed corporate 'personhood' to develop; that is, an abstract and disembodied 'legal person' without moral capacity.[56]

By contrast, Baruch Spinoza (incidentally born the same year as Locke) developed a radically different theory of personhood. Spinoza's theory was based on the idea of a single universal substance underlying the plurality of perceived differences; reality, Spinoza argued, is made of a single substance and the myriad of forms and ideas within the universe are but 'modes' of that unified substance. Spinoza's idea of an ultimate unity between God and Nature, or 'monism',[57] has been explored by Mathews,[58] Hampshire[59] and Gattens,[60] who have focused their attention on Spinoza's depiction of the self and its relationship within 'the whole'. Spinoza's hypothesis by necessity entailed individual responsibility toward the whole.[61] Similarly, Spinoza's concept of self did not divorce the 'mind' from 'nature', nor did it strip the world of its cosmological meanings and metaphysical character.[62] Spinoza did not espouse Descartes' mind/matter distinction, and the consequent 'mechanistic' concept of self, but conceived the self – or the body – 'to be a relatively complex individual, made up of a number of other bodies: its identity can never be viewed as a final or finished product... since it is a body that is in constant interchange with its environment.'[63] Spinoza conceived a three-tiered layer self: firstly, a self possessing individual interests; such an individual self is immersed in a larger self – that is, nature – in order to enable its own survival; finally, even this larger self in turn exists within an even larger self that is the universe.[64]

It is clear that had Spinoza's 'monism' become the foundational theory underpinning contemporary discourse, the perceived separation between the 'body' of law and the natural 'bodies' to which that abstract corpus refers would likely have lost its validity. Furthermore, the 'dephysicalisation' and disembodiment of law may not have developed in quite the way they did. Why, then, was Locke's theory more 'successful', why did it prove more foundational to the development of the abstraction, separation and 'dephysicalisation' of law?

Underlying the contemporary concept of self and its ontological framework is a teleological imperative, an imperative through which the dominant worldview is socially construed and instructed. Such an imperative can be gleaned through a Foucauldian analysis of power – that is, power as a dynamic entity that produces social, legal, scientific and political discourses that effect behavior 'at the capillaries of the social body' and that legitimize a dominant paradigm through its 'apparatuses of power', or the vehicles by which its continuance is promoted.[65] In light of such an analysis, the relationship of liberal philosophers, such as Locke and Hobbes, with positivism is explained by Mathews as 'de-metaphysicalising some of the central mysteries of scholastic thought, for example the notions of substance, cause, and real and nominal essence'.[66] Foucault's dynamic notion of power suggests that discourses that emerged to support the positivist paradigm provided ideological tools that were facilitative of a worldview that supported and enabled, above all, economic expansion. The combination of discourses of freedom with discourses of prosperity fostered an externalizing of the self through the accumulation of property, and articulated a value system that has promoted consumerism, expansion and capitalism since that time.[67] 'To be a person is to be a proprietor and also to be property – the property of oneself'.[68] Locke's philosophy of 'natural rights' both accommodated Christianity within a framework of positivism and materialism and reflected the dualistic notion of mind and body: mind, as the instrument of 'personal will', or the essence of the self, and the body, as its personal property. Together with the accumulation of goods and real property, the body and its property became a representation or extension of the individual's rights and the rights of personhood.[69]

It is through social and scientific discourses – and their oppositions – that a dynamic repertoire of ideas exists in which the dominant paradigm and its agendas are legitimized, normalized, supported and sustained in a social system.[70] Once legislative and legal mechanisms determine the hierarchy of values, the dominant paradigm's implied assumptions will favor those values that have become normalized. Subsequently, the law has the effect of reinforcing those values within the social framework. The law hardly changes without a plethora of co-existing discourses that supply the weight for that change to occur, but when it does the legal system plays the powerful role of normalization within the social body.[71] Discourses and belief systems are constructs not only formulated as a response to the human need to understand our 'self' and our relationship within the world, but as an active ingredient, or precursor, to how we – as humans – conceive 'ourselves' and act within the world in which we live. That is, not only as a response to being in the world, but as a teleological mechanism within which we continually construct meaning through systems of values and ideas in which we divest authority and significance in order to achieve not only an understanding of self, but some purpose as a society – or system – as a whole.[72] Consequently, we suggest that to embrace a novel ecological paradigm requires a move away from a subject-object ontological framework whereby the concept of self dominates, to a part-whole framework where a concept of intrinsically intertwined 'singularities' (as opposed to separate and autonomous 'selves') and a dialogue between individuality and mutuality of said singularities constitute a more appropriate approach.

Where a society has become fractured from its geographical roots and is instead driven by purportedly abstract economic imperatives, the 'organic' role of culture may be more difficult to see, yet it would be erroneous to suppose such a relationship does not, or could not, exist. What may be argued is that, where a culture through its discourses, myths or meta-physical understandings, is not reflecting the necessities or mechanisms for its survival, the culture is rather producing unsustainable practices, eventually destined to clash with (and ultimately lose against) the larger environmental imperatives within which that culture is inscribed. Mathews states that 'we become disconnected from Nature when culture malfunctions, that is, when our systems of abstract representation delivers up a misleading or erroneous picture of the world.'[73] We suggest that while there existed in the 17th Century a culturally construed 'economic imperative' fueled by the industrial revolution and the desire for a capitalist economic expansion, which promoted a correlating and cultural construct of 'self', the 21st Century faces vastly different challenges and the contemporary (and equally culturally contingent) 'ecological imperative' requires a construction of self consistent with that imperative. Optimistic-ally, it may be that the recent extension of 'legal personhood' to natural systems is an anthropomorphic correlative of recognizing living systems as 'selves'. Whether non-economic interests and non-human rights can be assimilated into the existing normative discourse, and by extension given substantial weight within judicial and legal decision-making, remains to be seen.

Ideas such as rights of nature are particularly important, more for the fact that they force us to re-interrogate our normative embodiment rather than for their prescriptive possibilities. By interrogating the notion of the self that underpins the current 'disembodied' representation of law in light of the suggestions advanced by Earth Jurisprudence, an alternate cosmology that takes a holistic view of our existence is instead proposed, a view that might be emancipated from the individualistic, material and mechanical assumptions of liberal ideology through its atomistic under-pinnings.[74] The current legal description of the autonomous self, almost synonymous with the philosophy of John Locke, and situated within a liberal paradigm, now needs to be changed to encompass a description of self within an interconnected world of living systems, in order to reprioritize the value and legal protection awarded to living systems. A re-embodied concept of self, as represented by practical gestures such as the articulation of rights of nature and nature's legal standing in a number of jurisdictions, then, is not simply one of a static or of an 'innate' self, but rather of a self that exhibits characteristics both of its 'nature' and its 'culture'. As well as a self that perceives itself as an emotional, imagining and discerning being, it is a self that is instinctually, physically, subcon-sciously and intuitively connected to, and part of, its ecosystem – an 'ecological' self.[75]

Notes

1   Alessandro completed his LLB/LLM in Law in Italy with a specialization in comparative law and legal anthropology. His thesis comprised a field research project on pre-Colombian family protocols in the Andes of Peru, Bolivia and Ecuador. Alessandro has been involved in Indigenous rights for over 15 years, by participating in and supporting the drafting of the UN Declaration on the Rights of Indigenous Peoples in Geneva. His PhD thesis, conducted at the University of Wollongong and completed in 2011, focused on native title and legal pluralism in the Illawarra. More recently, Alessandro began to explore the emerging discourse on rights of nature, Wild Law and Earth Juris-prudence. His main area of interest is the intersection between this emerging discourse and different legal ontologies, with a particular focus on Indigenous legal structures. Alessandro is one of the founding members both of the Global Alliance for the Rights of Nature and of the Australian Wild Law Alliance and he has contributed to establish the Earth Laws Network at Southern Cross University. Alessandro is currently working as an Associate Lecturer in the School of Law and Justice at Southern Cross University. His

main areas of research are legal anthropology, comparative law, legal theory, Indigenous rights and ecological jurisprudence.

2   Gabrielle O'Shannessy is a recent graduate of Southern Cross University. Her Honour's research focused on the relationship of a cultural construct of self with regards to law. Prior to studying law she taught music, jazz history and music business for over 17 years. Following from her broad arts background, her interest lay in how the embodiment of culture is achieved through dynamic relationships between culture and self. Her recent focus is on the ecological role culture plays in both the development of identity and the maintenance of living systems, and asks how law can be embodied in a dynamic culture that requires an ecological imperative in order to survive.

3   *Southern Pacific Co v Jensen* 244 US 205 (1917) at 222.

4   See J M Kelly, *A Short History of Western Legal Theory*, Oxford: Oxford University Press, 1992.

5   See MDA Freeman, *Lloyd's Introduction to Jurisprudence*, London: Sweet& Maxwell, 8th ed, 2008, pp. 386–91.

6   HLA Hart, 'Positivism and the Separation of Law and Morals', *Harvard Law Review* 71, 1958, 593, 601.

7   Hans Kelsen, *Pure Theory of Law*, trans Max Knight, Berkley, CA: University of California Press, 1967.

8   John Austin, *The Province of Jurisprudence Determined*, Indianapolis: Hackett, 3rd ed, 1998, first published 1832.

9   Cicero famously wrote: 'true law is right reason in agreement with nature'. Cicero, *De Legibus*, in Kelly, op. cit., p. 58. It is worth noting that the comments regarding 'natural law theories' refer both to classical and contemporary natural law theories. See Raymond Wacks, *Understanding Jurisprudence*, Oxford: Oxford University Press, 3rd ed, 2012, pp. 15–26.

10   Wesley Hohfeld, for example, argues that rights are the product of a relationship with law, without which there do not exist 'natural rights': 'a legal right is any theoretical advantage conferred by recognized legal rules'. See Wesley Hohfeld, 'Some Fundamental Legal Conceptions as Applied to Judicial Reasoning', *Yale Law Journal* 23, 1913, 16.

11   See Wesley Hohfeld, 'Fundamental Legal Conceptions as Applied in Judicial Reasoning', *Yale Law Journal* 26, 1917, 710; Ngaire Naffine and Rosemary J Owens (eds), *Sexing the Subject of Law*, 1996.

12   Nicole Graham, *Lawscape*, Oxon: Routledge, 2011, p. xiii.

13   A Pottage and M Mundy, *Law, anthropology, and the constitution of the social: making persons and things*, Cambridge: Cambridge University Press, 2004, p.3.

14   P Fitzpatrick, *The mythology of modern law*, London: Routledge, 1992.

15   Graham, op. cit., p. 27.

16   Ibid, p. 28.

17   Ibid, p. 23.

18   Ibid.

19   Swimme and Tucker, op. cit., pp. 99-102.

20   Oliver Wendell Holmes, *The Common Law*, Boston, MA: Little, Brown, 1881, p. 1.

21   Davies, op. cit., p. 159.

22   Ibid, p. 115.

23   See, for example, Gary Minda, *Postmodern Legal Movements*, New York: New York University Press, 1995.

24   Lawrence Rosen, *Law as Culture*, Princeton, NJ: Princeton University Press, 2006, p. 7.

25   Costas Douzinas and Adam Gearey, *Critical Jurisprudence*, Portland, OR: Hart, 2005, p. 47.

26   Michel Foucault, *The Order of Things* London: Tavistock, 1974.

27   Jacques Deridda, *Positions*, Chicago, IL: University of Chicago Press, 1981; J Deridda, *Writing and Difference*, Chicago, IL: University of Chicago Press, 1978; J Deridda, *On Grammatology*, Baltimore, MD: John Hopkins University, 1974.

28   Davies, op. cit., p. 217. See also Martha Fineman, 'Challenging Law, Establishing Differences: The Future of Feminist Legal Scholarship', *Florida Law Review* 42, 1990, 25; Catharine McKinnon, 'From Practice to Theory, or What is a White Woman Anyway?', *Yale Journal of Law and Feminism* 41, 1991, 13; Philippa Rothfield, 'Alternative Epistemologies, Politics and Feminism', *Social Analysis*, 30, 1991, 54.

29   Douzinas and Gearey, op. cit., p. 47.

30   Irene Watson, 'Indigenous Peoples' Law-Ways: Survival Against the Colonial State', *Australian Feminist Law Journal* 8, 1997, 39, p. 39.

31   Roderick Nash, *The Rights of Nature: A History of Environmental Ethics*, Madison, WI: University of Wisconsin Press, 1989.

32   Christopher Stone, 'Should Trees Have Standing? Towards Legal Rights for Natural Objects', *Southern California Law Review* 45, 1972, 150. See also Christopher Stone, *Should Trees Have Standing?*, New York: Oxford University Press, 3rd ed, 2010.

33   Thomas Berry, *The Dream of the Earth*, San Francisco, CA: Sierra Club Books, 1988; Thomas Berry, *The Great Work*, New York: Random House, 1999; Thomas Berry, *The Sacred Universe*, New York: Columbia University Press, 2009.

34   Cormac Cullinan, *Wild Law*, White River Junction: Chelsea Green Publishing, 2002.

35   Cormac Cullinan, 'A History of Wild Law', in Peter Burdon (ed.), *Exploring Wild Law*, Kent Town: Wakefield Press, 2011. Cullinan's argument articulates in legal theoretical terms the application of evolutionary theory to social organization offered by a number of authors. That is, the insight that human societies are exposed to the process of natural selection in relation to issues of long-term environmental sustainability just as single individual organisms are in relation to individual survival. Cullinan's argument advances the idea further, by suggesting that human societies determine – or at least influence to a significant degree – their path to sustainability – or lack thereof – through a specific set of legal structures that implement and normativize specific cultural worldviews. See Jared Diamond, *Guns, Germs and Steel*, London: Chatto & Windus, 1997; Jared Diamond, *Collapse*, New York: Viking Press, 2006; Edward O Wilson, *Sociobiology*, Cambridge: Harvard University Press, 1975; Edward O Wilson, *The Social Conquest of Earth*, New York: Liveright, 2012; Marvin Harris, *Cows, Pigs, Wars and Witches*, New York: Random House, 1974; Marvin Harris, *Cannibals and Kings*, New York: Random House, 1977; Joseph Tainter, *The Collapse of Complex Societies*, Cambridge: Cambridge University Press, 1988.

36   James Lovelock, *Gaia*, Oxford: Oxford University Press, 1979.
37   Cullinan, op. cit., Part IV. See also Tom Jagtenberg and David McKie, *Eco-Impacts and the Greening of Postmodernity*, Thousand Oaks, CA: Sage, 1996.
38   Cullinan, op. cit., pp. 59-61.
39   See Global Alliance for the Rights of Nature, http://therightsofnature.org.
40   See Peter Boulot and Helen Sungaila, 'A New Legal Paradigm: Towards a Jurisprudence Based on Ecological Sovereignty', *Macquarie Journal of International and Comparative Environmental Law* 8(1), 2012, 1.
41   Peter Burdon, 'The Great Jurisprudence' in Peter Burdon (ed.), *Exploring Wild Law. The Philosophy of Earth Jurisprudence*, Kent Town: Wakefield Press, 2011, p. 66.
42   Ibid, p. 68.
43   See Freeman, op. cit., p . 100.
44   Anne Schillmoller and Alessandro Pelizzon, 'Mapping the Terrain of Earth Jurisprudence: Landscape, Thresholds and Horizons', *Environmental and Earth Law Journal* 3(1), 2013 (forthcoming).
45   Ibid, p. 3.
46   Ibid, p. 11.
47   See Wacks, op. cit., pp. 39-40.
48   See Alessandro Pelizzon, 'Earth Laws, Rights of Nature and Legal Pluralism' in Michelle Maloney and Peter Burdon (eds), *Confronting Collapse: What Agencies, Institutions and Strategies Are Needed for a Better World? How to Achieve Environmental Justice?*, Routledge, 2013 (forthcoming).
49   Graham, op. cit., p. 23.
50   Margaret Davies and Ngaire Naffine, *Are Persons Property? Legal Debates about Property and Personhood*, Ashgate Dartmouth Publishing, 2001, p. 56.
51   Theodore Roszak, *The Voice of the Earth*, Bantam Press, 1993, p. 18.
52   Graham, op. cit., p. 48.
53   Pheng Cheah, David Fraser and Judith Grbich (eds), *Thinking Through the Body of the Law*, 1996, p. xiii.
54   Davies and Naffine, op. cit., pp. 1-15.
55   Ibid.
56   Aidan Ricketts, *Stretching the Metaphor: The Political Rights of the Corporate Person'* Master of Laws Thesis, Queensland University of Technology, 2001.
57   Ibid, p. 207.
58   Freya Matthews, *The Ecological Self*, London, Routledge, 1991, p. 28.
59   Stuart Hampshire, *Spinoza*, Penguin Books, 1951.
60   Gattens, op. cit..
61   Ibid, p. 30.
62   Stuart Hampshire, op. cit., p. 58.
63   Gattens, op. cit., p. 28.
64   Ibid, p. 29.
65   Foucault, op. cit..
66   Mathews, op. cit., p. 21.
67   Ibid, p. 36.
68   Davies and Naffine, op. cit., p. 5.
69   Ibid, p. 27.
70   Ibid, p. 127.

71   Cullinan, op. cit., p. 57.
72   Gordon, op. cit., p. 127.
73   Ibid.
74   See Mathews, op. cit., p. 5.
75   Ibid, p. 107.

Chapter 11

Beyond legal facts and discourses

Towards a social-ecological production of the legal

Margherita Pieraccini

Introduction

Legal pluralism is an attractive concept for use in studying multi-layered legal fields. This is because legal pluralism puts the accent on legal complexity, legal relationality and legal evolution or dynamism. Indeed, legal pluralism describes how societies contain multiple normative orders and discourses of the law, some of which are produced and maintained by institutions other than the state (i.e. legal complexity). Such pluralism looks at the relationships between state law and unofficial forms of law (i.e. laws not recognized by the official authorities of a state, but recognized by the practices and consensus of determined communities) and between law and non-legal orderings (i.e. different types of power relationships and interpenetrations between legal spheres) and it considers the extent to which law is able to respond to societal complexity through time (i.e. how what counts as law in a particular environment evolves due to new societal sensibilities).

The question addressed by this chapter concerns the extent to which understandings of legal pluralism are able to capture the complexity of systems dynamically co-constituted by social and ecological elements. Much depends on the understanding of law offered by legal pluralism and in this chapter I identify two distinct streams – one (classical legal pluralism), which relies on an understanding of law as a social phenomenon/fact, a visible occurrence to be studied empirically. This stream focuses on the existence of a plurality of legal orders and pays attention to their social-political effects. Legal pluralism in this sense is an observable occurrence, understood in terms of legal dynamism, human agency and power relations.

I also identify a second stream, i.e. the discursive approach to legal pluralism, which is not as concerned with finding legal plurality in observable domains of social control, but more with the discursive interpretations of what constitutes law. This is especially visible in autopoietic accounts of systems' interactions. Autopoietic accounts pay full attention to linguistic structures of signification of the legal by identifying systems using the binary code legal/non-legal, positioning at the centre of the analysis a description of systemic self-production and (mis-)communication. Though very different, all these legal pluralist approaches portray law as a social construction, whether as a social fact or a discourse.

After identifying these two distinct streams of legal pluralism, I argue there may be a third way in which we can think about legal pluralism that entails a greater attentiveness to social-ecological engagements. Differently from the two categories of legal pluralism around which much of the scholarship has mushroomed, this third perspective, primarily influenced by law and geography scholarship, pays attention to the way in which law is performatively[1] given meanings in social-ecological settings, the way in which the law is produced by a process involving the practices and bodies of human and non-human entities. Rather than focusing on social phenomena or systemically ascribing legal meanings, this third approach revives the world of social-environmental practice and emphasizes the importance of re-embodying the law and accounting for social-ecological relationality within legal pluralist analysis. This third strand of legal pluralism I name social-ecological legal pluralism, and two examples are offered below to ground the theoretical insights in empirical realities.

Part I: legal pluralism as a fully social construction

This part of this chapter highlights established perspectives in the legal pluralist scholarship subdivided into two streams: classical legal pluralism, where law is an empirical social fact, and autopoietic understandings, where law is a discourse. There are substantial differences between these perspectives though they have a common core: a conception of legal pluralism as a fully social-discursive occurrence, where social-ecological interactions in the production of the legal are little – if at all – contemplated.

At the core of legal pluralism lies an understanding of law as a complex and varied occurrence. Legal pluralists have opened up law to embrace different types of legal orders, not necessarily tied to the state machinery. A well-known definition proposed by Griffiths states that legal pluralism is 'the presence in a social field of more than one legal order.'[2]

This flexible understanding of what counts as law can be traced back to colonial times, where a number of interesting ethnographies contained accounts of indigenous law, asking the question of how people maintained social order in 'stateless society'. 'Folk'/'primitive law' became a category of inquiry as anthropologists recognized that small-scale societies, in the absence of specific juridical institutions, were able to maintain order and comply with socially approved norms. Initially, these anthropologists documented the inner workings of 'primitive law'.[3] These early studies differ between a more process-oriented/functionalist understanding of law (represented by Malinowski, who documented the practical ways in which reciprocal economic obligations between coast and mountain Melanesian people ensured the social order of the Trobriand society)[4] and a structuralist one (drawing on Durkheim, and visible in Radcliffe-Brown's work).[5] These anthropologists had different ways of conceiving and defining law: some, like Radcliffe-Brown, considered law as an institution of control exercized through force in politically organized societies studied as an isolated subject while others, like Malinowski, proposed a looser definition of law as social control; some, like Gluckman,[6] used the lexicon of Western law to describe rules and disputes in non-Western societies; others, like Bohannan,[7] considered this application to be inappropriate and pushed towards the use of indigenous terms to describe non-Western law. However, all of these authors tended to discuss 'primitive law' by itself and in itself without considering the larger colonial context in which it had become encapsulated.

It was only from the 1970s that anthropological and historical attention shifted towards the analysis of the encounters between colonial rule and 'primitive law', pointing to the porosity of law (the borrowing of a system from another) as well as to power relations between the two systems. It is here that legal pluralism in its modern form began to emerge. Exploring the dual legal systems at play in colonial settings meant examining the extent to which customary law was subsumed under, transformed by and, sometimes transforming, colonial law. The matter of power became central to legal pluralism. For instance, Ranger, in discussing the invention of neo-traditions in colonial Africa, spoke of the subordination and rigidification of evolving and dynamic local customs to fixed categories and to colonial rule: 'what we called customary law, customary land-rights, customary political structures and so on, were in fact *all* invented by colonial codification... once the 'traditions', relating to community identity and land rights were written down in court records and exposed to the criteria of the invented customary model, a new and unchanging body of tradition had been created.'[8] Ranger documented the way in which the invented customary law served as a mechanism for power not only to colonial rulers but also to some sections of the African societies (elders, men, ruling aristocracies and indigenous groups) to assert their dominance of rural means of production, of political and economic rights; 'codified and reified custom was manipulated by such vested interests as a means of asserting or increasing control'.[9]

If Ranger and others[10] conceptualized a break between pre-colonial and the 'invented' colonial legal worlds, post-modern scholars, such as de Sousa Santos, have documented a more complex power relationship between legal orders in colonized societies pointing to the hybridization, cross-fertilization and cross-contamination between legal boundaries.[11] For de Sousa

Santos, 'legal pluralism is the key concept in a post-modern view of law. Not the legal pluralism of traditional legal anthropology in which the different legal orders are conceived as separate entities coexisting in the same political space, but rather the conception of different legal spaces super-imposed, interpenetrated, and mixed in our minds as much as in our actions... we live in a time of porous legality or of legal porosity, of multiple networks of legal orders forcing us to constant transitions and trespassings. Our legal life is constituted by an intersection of different legal orders, that is, by interlegality.'[12]

Using the metaphor of law as a map and relying on geographical con-cepts, de Sousa Santos[13] presents a symbolic cartography of law, arguing that law is a social construction of reality dependent on scale, projection and symbolization. As explained by de Sousa Santos, the larger the scale, the finer are the details represented and vice versa, so that domestic law tends to be more accurate than international law in terms of what gets represented but it is less able to provide a general framework for orienting the legal subject. Projection refers to the definition and organization of the legal space and symbolization on the formalization of the everyday in selected abstractions. Due to a variety of not necessarily synchronic scales, projections and symbolizations operate and inter-penetrate, we have a situation of legal pluralism.

For Melissaris, de Sousa Santos' work begins to shift the legal pluralist's focus 'from the recognition of *legal systems* reducible to certain criteria to *the relations developing between dispersed legalities, discourses of a legal quality*'.[14] 'Prudent knowledge' becomes a major component in understanding plural constellations of law in post-modern times as attention is put on the episte-mological forms of legal systems.[15] In my view, however, the writing of de Sousa Santos does not constitute such a break with the first stream of legal pluralism identified above as they are still rooted in the empirical tradition of the first strand of legal pluralism. Twining himself includes de Sousa Santos in what he identifies as a 'social fact' view of legal pluralism.[16] Indeed, legal orders exist out there for Santos in six social landscapes: household-place, work-place, market-place, community-place, citizen-place and world-place.

The understanding of law as a discourse is, for me, much more visible in autopoietic accounts. If the rejection of positivist accounts of law is a common trait between the work of post-modern legal pluralist scholars, such as de Sousa Santos and of autopoietic scholars, autopoietic accounts go a step further in defining law as communication. For Teubner, legal pluralism is defined 'no longer as a set of conflicting social norms in a given social field but as a multiplicity of diverse communicative processes that observe social action under the binary code of legal/illegal'.[17] Law is to be found whenever the code legal/illegal is employed. In this way, Teubner attempts to overcome one of the main critiques raised against legal pluralist scholars: i.e. the stretching out of law to every field of social control, thereby risking conflating the term law with social norm.[18] Such a critique can be raised against much legal pluralist scholarship summarized above, even the work of de Sousa Santos, since the six social landscapes in the *Towards a Legal Common Sense* position law everywhere. The boundary between law and social life becomes blurred, or better, there is a juridification of most spheres of the social world.[19]

The discursive approach to legal pluralism advocated by Teubner is immune to this criticism because it is not concerned with finding legal plurality in observable domains of social control but in distinguishable discursive interpretations of what constitutes law. Legal pluralism consists in recognizing the existence of a plurality of regulatory discourses that use the binary code legal/illegal. Teubner argues that the legal should not be defined according to function (anything that serves the function of social control) or structure (any normative expectations) but according to the binary code employed.[20] All phenomena communicatively observed using the binary code legal/illegal are within the legal pluralist set. This code is not confined to the law of the State, as the juridification of social phenomena does not only happen when the state recognizes them as 'law'. This allows Teubner to make a sharper distinction between law and other forms of social control, without falling into the fallacy of 'legal centralism'. The image of legal pluralism proposed by Teubner is not hierarchical with the state at the apex, but heterarchical: a plurality of fragmented legal discourses occurring simultaneously. The relationship between the legal and non-legal systems is to be understood via the concepts of structural couplings and productive misreading. Structural couplings between systems emerge from the cognitive openness of system, not their normative openness, thereby leading to 'productive misreadings', i.e. the legal misreading of non-legal social discourses as sources of norm production.[21]

The analytical focus therefore is on the discursive invocation of the legal code in various social spheres, on the 'recording of a bewildering multitude of otherwise coded communication in the code of law'.[22] However, as Tamanah observes, 'characterizing law exclusively in terms of communication loses direct touch with the material power and effects of law'.[23] The autopoietic version of legal pluralism indeed de-materializes law completely as it collocates it in the realm of systemic communication, reducing law to pure signification. If the first (classical) stream of legal pluralism hovers over the material world concentrating on forms of social control, the autopoietic version distances itself even more from materiality because discourses, rather than practices, become central.

The discursive turn does not mean that law as a social phenomenon, as the existence of more than one visible normative order in the same social field, is now a forgotten item of the legal pluralist agenda. Indeed, the interactions between normative fields are still studied today but the geographical scope has been enlarged so that, though legal pluralism in the former colonies remains an important analytical focus, legal pluralism as a social phenomenon is documented outside colonial society also as academics recognize that the condition of legal pluralism is endemic to every society.[24] The work of Moore on 'semi-autonomous' social fields, akin to Malinowski's processual approach to law, was pioneering in this respect for considering 'the internal workings of an observable social field and its points of articulation with a larger setting' and in so-called complex societies (the dress industry in New York was given as an example by Moore).[25] Legal pluralism 'at home' has since become an important subject of study by itself or as part of comparative inquiries, which consider legal similarities and differences between societies and through time.

Besides this, a number of current writings transcend single societies 'transnationalizing' and globalizing legal pluralism.[26] The 'transnationalization' of legal pluralism does not only occur in classical legal pluralists' accounts, which relate legal pluralism to identifiable empirical legal fields, but also in those following the discursive turn. Again, the work of Teubner can be cited. In 'Global Bukowina', Teubner gives the example of *lex mercatoria* to explain global legal pluralism.[27] The insulation of *lex mercatoria* from the state reflects the legal pluralist's rejection of legal centralism, and highly technical, specialized economic global networks render *lex mercatoria* a self-reproducing system in structural couplings with global economic institutions and transactions.

Despite their marked differences, the established approaches to legal pluralism discussed above share an understanding of the making and the production of law as an exclusively social achievement: actors, institutions and discourses are fully social constructions. The material world is somewhat in the background as a passive surface over which legal pluralism unfolds. None of these accounts pay adequate attention to the way in which law is the co-production of social and ecological environments, the way in which law takes form through bodies and practices of human and non-human animals as much as through discourses or social facts. In my view, much could be learnt from social science efforts to 'rematerialize' and re-embody the social world – and more specifically from law and geography scholars who in the last decades have translated these observations into the study of law, condemning a reified understanding of law, replacing it with insights pointing to the relational entanglements between law and space and the spatial contingency of legal interpretations and legal occurrences.

As will be recalled, de Sousa Santos also employed the spatial metaphor to discuss the scales of law, and the first strand of legal pluralism pointed to the importance of empirical studies to find what law meant in particular localities and peripheries. However, there is a fundamental difference between the two approaches. As Manderson puts it, what law and geography scholarship adds to the picture is the exploration of 'the diversity of legal norms and the disparateness of legal effect not just in terms of the social elements that constantly work to generate and differentiate it, but the physical elements too and of course the social and the physical are likewise mutually implicated.'[28] Whilst de Sousa Santos does not move beyond symbolism and discourses, some critical legal geographers take the material world seriously. As will be shown in the next part of the chapter, some insights of law and geography scholarship have entered legal pluralist accounts but the legal pluralist literature that embraces the spatial turn is still thin and there has not been any attempt to conceptualize this as a new shift: a new legal pluralism. The purpose of the following part of this chapter is twofold: first of all, to introduce the literature of critical social sciences and law and geography scholarship, in particular on re-materialization and, second, to provide examples of legal pluralist scholarship moving in this direction in order to concretize this new orientation (the third way) of approaching and making sense of the legal.

Part II: towards a third way in legal pluralism

In recent decades, various currents of critical social sciences have moved away from the interpretative and discursive turns pointing to the connections between *bio* and *geo*. As Whatmore argues,[29] these currents can be best characterized as 'materialist returns' because in constructing their arguments, they draw on philosophical insights of Deleuze and Guattari and other thinkers (from Gibson to Merleau-Ponty to Latour) by reanimating matter in describing social life. These 'materialist recuperations', as Whatmore calls them, include work by social anthropologists and cultural geographers.

At the core of these accounts lie a rejection of particular Western philosophical influences of understanding space, rooted in the Cartesian dichotomy between *res cogitans* (i.e. mental substance) and the *res extensa* (i.e. extended thing) and in the Kantian consideration of space as *a priori* category, existing in the realm of consciousness.[30] The rejection of this tradition is evident in Ingold's critique of the 'global' way of understanding the earth as opposed to the 'spherical' one, with the former conceiving the environment as a place we look at and appropriate rather than a place we inhabit[31] or in Whatmore's call for a 'more-than-human geography'[32] resting on a recognition of the multiple interweaving of material and social life, or again in Bennet's 'thing-power'[33] and 'vibrant matter',[34] concepts used to emphasize the way human culture is enmeshed with the agency of materials and its environmental ethics and democratic repercussions.[35] The list could continue citing the work of Escobar on the transition towards the 'pluriverse' as understood as 'an ever-changing web of interconnections involving humans and non-humans'[36] and Descola's *Beyond Nature and Culture*[37] in which the author attempts to overcome the universalizing provincialism of the Euro-American tradition centred on a strict separation between the cultural worlds of human beings and the material world of non-human nature.

Through these and many other accounts, we are exposed to a shift towards relational ontologies, advocating that much of the social is produced as part of its entanglement with the material. These insights have somewhat entered the study of law in the field of law and geography. Law and geography scholars have pointed out how law and space are not objective, separate categories; rather that they are recursively created with law constituting and being constituted by space.[38] As Holder and Harrison note,[39] uncovering the links between law and space has 'the capacity to release law from its (imposed and self-imposed) confinement as 'word' (interpretation, meaning, discourse) and to show 'the legal at work in the world', a world dominated by the physical (places, landscapes, spatialities, natures)'.

The spatialization of law and the legalization of space have been well captured by the work of Blomley in the last two decades. Blomley argues against the 'tendency to focus on representation, cultural and discourse turning away the attention from the world of things'[40] because this silences the significance and effects of things and their entanglements with different legal understandings. His work on natural features acting as boundary-makers for property relations contains important insights for the law and geography project. Blomley shows how using nature as a property-boundary maker is complicated, even when it is a nature that apparently lends itself to this task, such as a river or a hedge. For example, in re-reading the dispute over the Blackbird Bend of the Missouri river between the Omanha Indian Tribe and the state of Iowa, Blomley[41] uncovers the complex interrelationships and mutual constitution between the law (or, preferably, laws) of property and the river. Although the court recognized nature, tracing the movement of the river and pointing to its wild status, it did so discerning categories that are tied to particular social concepts of property and that proved difficult to be reconciled with the unpredictable flows of the river.

Similarly, in discussing the enclosure movement in early modern England, Blomley again points to the role played by material things (hedges in this case) in making, enforcing and at times disrupting (because their materiality made them also subject to breaking) new forms of law and social relations.[42] The point Blomley advanced in this paper is again that historians and other scholars of enclosure have overlooked the material world in answering the question of 'how did enclosure happen?' by focusing only on discourses and representations, as manifested in scholarly attention on maps as a powerful representational tool. Blomley argues that hedges instead played an important role in the making of private property and in facilitating enclosure. The hedge indeed served to enclose and to signify private property and, by restricting human and animal movements on the land, contributed to defend this newly created private property from commoners. Therefore, rather than solely being a material product of the *a priori* legal and political project of enclosure, hedges contributed to its making actively, as a sign of private property and as a tool to enforce it. In the words of Blomley 'the hedge also reminds us of the more general importance of material things to property. For all sorts of reasons, property (both in law and scholarship) has tended to eschew the material and the embodied. Property is about relations and representations, it is argued. But property is also productive of, and reliant upon, things'.[43]

To signify the enmeshing of law and geography, the simultaneous legal and spatial essence of particular orderings, Blomley speaks of 'splices' and 'splicing'.[44] Examples of splices are a citizen, a legal category that necessitates the spatial category of a territory to make sense; or a prisoner who needs the spatial category of a prison to exist as such. The splices are particular categories where law and space merge. However, although they appear as given, natural and stable, the splices are not fixed, given orderings because they are the product of on-going enactments, through acts of splicing. This confers on them a processual and dynamic aspect: 'it helps if we recognize dominant splices as splicings; as achievements rather than facts'.[45] To illustrate the meaning of 'splicing', Blomley offers the example of private property that requires a number of 'sustained enactments', such as cadastral mapping; everyday practices by home owners to maintain the boundaries between their property and that of others; and formal legal channels sanctioning private property and discourses related to property. He also states that 'it is important to remember that it is not just law that is enacted. In enacting law, we enact space, and vice versa. A cadastral grid, of course, exists only to the extent that is internalized and put to work by people in the everyday'.[46] If splices are understood not as nouns but as verbs, as splicings, they can also be disrupted and reshaped. If so, re-splicing becomes also possible, i.e. forms of resistance against the naturalization of these categories, their deconstruction and formation of alternatives.

Delaney has taken the splicing metaphor further, with the neologism 'nomosphere.'[47] In Delaney's words, the nomosphere refers to 'cultural-material environs that are constituted by the reciprocal materialization of the 'legal', and the legal signification of the 'social-political', and the practical, performative engagements through which such constitutive movements happen and unfold'.[48] If splices emphasize the co-constitution between law and space, the nomosphere (i.e. the sphere of law) is about considering how the entanglements between the material and the discursive play out in social practices and have repercussions on the lived experience of the subject. Both Delaney and Blomley argue against a dematerialization of the legal, advocating instead analyses of legal worldliness, i.e. how law manifests itself in and through the material world. De-materializing the law by concentrating solely on discourses, words and framings is, for Delaney, a political exercise in purification, boundary-work aimed at distancing law from its context. And this purification goes against a syllogism that Delaney advances, drawing on Actor Network Theory's insights on the mutual constitution of the social and the material: 'if the legal is constitutive of the social, if the social is irreducibly material-in a non-reductionist sense – and if we carry the social in our very bodies, then there is no a priori way of dematerializing the legal'.[49] Prisons present a paradigmatic case of legal materialization, as they are legally defined spaces to act on the bodies of the inmates, for example by injecting pharmaceuticals to moderate their behaviour. And it is not only through acting on subjects that law is materialized in prisons but also through things such as doors, cells, locks and files that constitute the material embodiment of the legal.

These insights are very interesting in pointing to the material manifestations of the legal and vice versa and therefore are helpful for rethinking how law matters. However, these accounts do not explicitly engage with legal pluralism. They are interested in a particular manifestation (be it the hedge or the prison) of the relationship between word and deed, discourse and material and law and space rather than in situations of legal pluralism.

As Benda-Beckmann et al. argue in the edited collection *Spatializing Law*[50] the literature on law and space has so far not paid much attention to situations of legal pluralism. According to Benda-Beckmann at al., this is due to the fact that most of the law and geography literature has focused on state law in industrialized states, and has an urban bias. However, the authors argue, many people live under legal pluralist constellations and therefore there are different constructions of space in one setting, produced by the state, custom and religion operating simultaneously and often in conflict with one another. For Benda-Beckmann *et al.* there are a number of fully-socially produced legal orders that construct space in different ways and raise competing claims over resources, opening up arenas for the exercise of power. Second, the authors argue that the negotiation between different legally relevant notions of space is primarily visible in post-colonial societies. Many chapters of their edited collection show how state constructions of space draw new criteria to administer natural resources, undermining local constructions of space. For example, Wiber shows this in relation to Canadian fisheries management in the Bay of Fundy, arguing that administrative law has developed technocratic management regimes that undermine practice-based natural resources' knowledge of local fishers and regulate the fish stocks at a different scale than before.[51]

Another more recent example, this time of the internal pluralism of state law, is brought by the analysis of competing legalities on Mt Taylor in the American Far West discussed by Harm Benson.[52] As Mt Taylor is public land, uranium mining can take place under the General Mining Law 1872, which provides a right of entry, extraction and sale of the uranium to the miners. At the same time, Mt Taylor is also sacred land for many indigenous groups so that it has acquired the status of traditional cultural property under the National Historic Preservation Act 1966. According to Harm Benson,[53] the General Mining Act 1872 and the National Historic Preservation Act 1966 enact two different ontologies: the first enacts Western ideas of private ownership and control over nature (the mountain as a space to be appropriate for uranium mining purposes within a framework of private ownership), while the latter enacts indigenous peoples' ways of relating to the mountain. The two laws (the 1872 Act and the 1966 Act) serve to represent contrasting ways in which peoples make sense of and relate to the mountain landscape.

These examples show some of the links between legal pluralism and law and geography scholarship by considering how different legalities raise and enact competing different legal and moral claims about space and resources. In the remainder of this chapter, however, I would like to show another type of link between legal pluralism and law and geography scholarship by asking what is added to the legal pluralist agenda if we take the insights of the co-creation of the material and the social seriously. What does the social-ecological turn mean for legal pluralism and what type of legal plurality do legal pluralist researchers need to search for? The emphasis here is not on how law and geography literature could benefit from considering societies that live under legal pluralist constellations, but rather how the legal pluralist scholarship could benefit from considering the legal also from a social-ecological viewpoint. So my concern is more with different productions of the legal than with the social constructions of space in different legalities. To explore this, the next section offers two examples of the social-ecological production of the law in legal pluralist settings.

Two examples of socially ecologically produced legal pluralism

Uprooting Palestinian olive trees and sheep grazing on upland English and Welsh commons seem to have very little in common, but it will be shown that an analysis of these activities requires an appreciation of materiality for understanding the legal pluralism they generate. These two examples come from the work of Bravermen in Palestine[54] and from my own work in England and Wales.[55] By no means are these the only examples of a materialization of legal pluralism,[56] but they are sufficiently diverse contextually to be able to offer a wide-enough array of cases and to concretize the argument.

The olive tree, Bravermen tells us, has become the embodiment of Palestinian nationhood and resistance to Israel's occupation as the olive tree's steadfastness (*sumud*) is used to strengthen territorial claims of the Palestinians.[57] This is because the 'ownership' of the tree implies the 'ownership' of the land. Olive trees have however been uprooted not only by the state of Israel to make space for the Separation Barrier but also by the sabotaging acts of some Jewish settlers. However, interestingly, some human rights groups supporting Palestinian farmers are replanting trees not for environmental reasons but to re-establish a claim to the land. This is a first example of the way the tree enters the realm of legality: it is a politico-legal means to stop the unjust appropriation of Palestinian land, to reclaim property in land.

The act of sabotaging trees by Israel settlers is deemed illegal by the Israel state and consequently Israel has developed a system of administration related to Palestinian access to land in order to minimize disruption of Palestinian harvest by settlers. However, this system, which required Palestinians to initiate requests for protection, was challenged on human rights grounds in the Israeli Supreme Court that ruled against the Israeli policy requiring Israel's Defence Force to support Palestinian olive cultivation. The Israel State had established a complex bureaucratic system consisting of a calendar to harvest in particular days and in new spatial delimitations, reducing the landscape 'to an abstract, two-dimensional series of blue, red, and yellow zones that have come to represent regimented codes of mandatory spatial conduct'.[58] These new layers of legality have affected the Palestinian relationship with the olive tree and the land. The uprooting of the olive tree therefore contributes to the production of plural legalities regarding the scope of property rights: through uprooting, new Israeli settlers attempt to disrupt the relationship of Palestinian people to their land; through replanting the tree, human rights associations attempt to make new ownership claims to the land; and through the protection of Palestinian olive trees via a tight administrative system, the Israeli Defence Force again re-shapes the right to land of the Palestinians, strengthening Israeli control over Palestinian movements in the West Bank. It is not the olive tree in itself that generates property pluralism but the performative acts of uprooting and replanting the tree and limiting access to its harvest that shows how legal pluralism takes place in this particular context.

I now consider English and Welsh common land, an example I take from my own research. Common land is mostly privately owned land where commoners (i.e. third parties) have rights of common (i.e. use rights) over it. On upland commons the rights of pasture (i.e. rights to graze livestock) are still an important element of the commoners' farming economy. Part I of the Commons Act 2006 regulates the registration of rights of common in commons registers. The registers should specify the nature of the right, whether it is attached to a dominant tenement (a landholding near the common), and, if not, the name of the owner of the right. Upland commoners can therefore exercise the registered rights[59] on the common. The statutory law does not impose spatial limitations for the exercise of rights of pasture within a common, which remains a shared unfenced pasture. But to what extent are upland commons truly shared? Although the rights of common can be exercised on the whole common, rather than on a particular area of it, the heafting system in England and the Welsh equivalent (the sheepwalks) mean that in practice the common is subdivided into distinctive pockets of land that sometimes commoners refer to with the language of property. A heaf is a unit of land to which a flock of sheep learns to attach itself so the common is divided in distinct sections, each 'belonging' to a commoner because his/her flock of sheep are attached to it. The commoners' recognition of the settling behaviours of the sheep therefore determine property relationships that are important on common land. The sense of 'property' deriving from the flock's movement, from the bodies of the sheep on the land is even more pronounced in Wales, where the claim of a sheepwalk as the grazier's freehold mountain land has been tested, albeit unsuccessfully, in courts in the nineteenth century.[60] Still today, some Welsh commoners attempt to claim the sheepwalk as private mountain land and as a recognition of customary relationships. So, for instance, there have been claims raised against common registration authorities concerning the opposition to the registration of land as 'common' in the registers due to the existence of 'private' sheepwalks. These claims have however been rejected because the official legal view of common land – as a common space over which commoners have rights of commons – has been asserted. In sum, diverse and conflicting conceptions of property exist on common land: those deriving from common law and current statutory law and those more embodied ones deriving from social-ecological relations of sheep, land and farmers on the land.

Conclusion

This chapter has expanded the analytical focus of legal pluralism by arguing that human engagements with non-human subjects (and vice versa) play an active role in the construction of the legal. This third way of understanding legal pluralism emphasizes the social-ecological production of the legal. If the shift from law as fact to law as discourse has been explicitly advocated in the autopoietic work on legal pluralism, the shift from law as discourse to law as social-ecological practice has yet to be acknowledged, although it stems naturally from insights of law and geography scholarship and is of obvious relevance for analyses of legal pluralism related to environmental matters. Law and geography scholarship has in fact emphasized for the last two decades law's 'worldiness',[61] i.e. documenting legal inscriptions in lived environments. Moreover, some of the most critical law and geography scholars have put the accent on legal-ecological relationality through concepts, such as splicing or nomosphere. However, these accounts have not considered the implications of their ontological reflections for legal pluralism and this is what I have attempted to do here. Much like law and geography scholars, drawing on the examples of the olive tree and sheep grazing above, I attempted to show how the legal can also be produced by social-ecological encounters, whether by uprooting a tree or by the farmers' attention to sheep movements on the land. But this is not where the analysis should stop, because my aim has not only been to acknowledge this type of social-ecological legality but also to show there may be different ways in which this legality is manifested (the olive tree example) or that the social-ecological form of legality may contrast with other forms of legalities (the sheep example), thereby generating a situation of legal pluralism. This means that social-ecological law is not the only the form of law in a particular context but co-exists with other forms of legalities and normative orders, some solely socially produced. Also, being the contingent product of social-ecological relations, this legality is not fixed but is subject to change and re-workings. Such a call for acknowledging the social-ecological production of the legal should not be confounded with materialist reductionism, where the 'social' has been replaced by the 'thing' in an attempt to overcome social constructivist views, but should be understood as an effort to put firmly onto the legal pluralist agenda an analysis of performative social-ecological acts of making the legal and the way these can take multiple forms and might conflict with other forms of recognized law. If attention to the complexity, dynamism and relationality of the legal is a key feature of legal pluralism in its classical form, legal pluralist literature has primarily been interested in law as a social fact or as a discourse, but here it has been argued that a new turn should be recognized, a turn towards the social-ecological relationality of legal pluralism.

Notes

1   'The move toward performative alternatives to representationalism shifts the focus from questions of correspondence between descriptions and reality (e.g., do they mirror nature or culture?) to matters of practices/doings/actions.' K. Barad, 'Posthumanist Performativity: Toward an Understanding of How Matter Comes to Matter', *Signs*, 2003, vol. 28 (3), 801–31, p. 802. For discussions on performativity, see also A. Pickering, *The Mangle of Practice: Time, Agency and Science*, 1995, Chicago, IL: University of Chicago Press and J. Butler, *Bodies that Matter. On the Discursive Limits of Sex*, 1993, London and New York: Routledge.

2   J. Griffiths, 'What is legal pluralism?', *Journal of Law and Legal Pluralism*, 1986, vol. 24, 1–55.

3 E.g. B. Malinowski, *Crime and Custom in Savage Society*, 1926, New York: Hartcourt; A. R. Radcliffe-Brown, *The Andaman Islanders: a study in social anthropology*, 1922, Cambridge: Cambridge University Press and A. R. Radcliffe-Brown, *Structure and Function in Primitive Society*, 1952, Glencoe: the Free Press.

4 B. Malinowski, *Crime and Custom in Savage Society*, 1926, New York: Hartcourt.

5 J. L. Comaroff and S. A. Roberts, *Rules and Processes: the cultural logic of dispute in an African context.* 1981, Chicago, IL: Chicago University Press.

6 M. Gluckman, *The Judicial Process Among the Barotse of Northern Rhodesia*, 1955, Manchester: Manchester University Press.

7 P. Bohannan, *Justice and Judgment among the Tiv*, 1957, London: Oxford University Press for the International African Institute.

8 T. Ranger, 'The Invention of Tradition in Colonial Africa' in E. Hobsbawm and T. Ranger, *The Invention of Tradition*, 1983, Cambridge: Cambridge University Press, 211–62, pp. 250–51.

9 Ibid., p. 254.

10 E.g. M. Chanock, *Law, Custom and Social Order*, 1985, Cambridge: Cambridge University Press.

11 B. de Sousa Santos, 'The Heterogeneous State and Legal Pluralism in Mozambique', *Law & Society Review*, 2006, 40, 39–75.

12 B de Sousa Santos, 'Law: A Map of Misreading. Toward a Post-Modern Conception of Law', *Journal of Law and Society*, 1987, vol. 14, 279–302, pp. 297–298.

13 Ibid.

14 E Melissaris, 'The more the merrier? A new take on legal pluralism', *Social and Legal Studies*, 2004, 13: 57–79, p. 72, emphasis in the original .

15 B de Sousa Santos, *Toward a New Common Sense: Law, Science and Politics in the Paradigmatic Transition*, 1995, New York: Routledge, p. 489.

16 W Twining, 'Normative and Legal Pluralism: A Global Perspective', *Duke Journal of Comparative and International Law*, 2009, vol. 20, 473–517.

17 G Teubner, 'The two faces of Janus: Rethinking Legal Pluralism, *Cardozo Law Review*, 1992, vol. 13, 1443–1462, p. 1451.

18 B Tamanah, 'The Folly of the Social-Scientific Concept of Legal Pluralism', *Journal of Law and Society*, vol. 20, 192–217.

19 Above n. 14.

20 G Teubner, 'Global Bukowina: Legal Pluralism in the World Society' in G Teubner (ed.) Global Law without a State. Brookfield: Darmouth, pp. 3–28.

21 Ibid, p. 1447.

22 Ibid, p. 1457.

23 B Tamanah, 'A non-essentialist version of legal pluralism', *Journal of Law and Society*, vol. 27, 296–321, p. 311.

24 S E Merry, 'Legal Pluralism', *Law and Society Review*, 1988, vol. 22, 869–896.

25 S F Moore, *Law as Process: an anthropological approach*, 1978, London: Routledge.

26 E.g. S E Merry, International law and sociolegal scholarship: toward a spatial global legal pluralism, *Stud. Law Polit. Soc.*, 2008, vol 41, 149–168; R Michaels, Global Legal Pluralism, *Annual Review of Law and Social Science*, 2009, Vol. 5, 243–262.

27 Above, n. 20.

28   D Manderson, 'Interstices: New Work on Legal Spaces', *Law Text Culture*, 2005, vol. 9, 1–10, p. 1.

29   S Whatmore, 'Materilist returns: practising cultural geography in and for a more-than-human world' *Cultural Geographies*, 2006, vol. 13, 600–609.

30   C Butler, 'Critical Legal Studies and the Politics of Space', *Social and Legal Studies*, 2009, vol. 18, 313–332.

31   T Ingold, *The Perception of the Environment: Essays on Livelihoods, Dwelling and Skills*, London: Routledge, Ch. 12. The point of inhabiting is also expressed by Ingold more recently: 'we inhabit our environment: we are part of it; and through this practice of habitation it becomes part of us'. T Ingold, *Being Alive: Essays on Movement, Knowledge and Description*, 2010, London: Routledge, p. 95.

32   S Whatmore, *Hybrid Geographies: Natures Cultures Spaces*, 2002, London: Sage. The proposal to use the term 'more-than-human' came originally from the philosopher Abram. See, D Abram, *The Spell of the Sensuous*, 1996, New York: Vintage Books.

33   J Bennett, 'The Force of Things: Steps toward an Ecology of Matter', *Political Theory*, 2004, vol. 32, pp. 347–372.

34   J Bennett, *Vibrant matter: a political ecology of things*, 2010, Duke University, NC: Duke University Press.

35   There is 'an affinity between thing-power materialism and ecological thinking: both advocate the cultivation of an enhanced sense of the extent to which all things are spun together in a dense web, and both warn of the self-destructive character of human actions that are reckless with regard to other nodes of the web'. Ibid, p. 354.

36   A Escobar, 'Notes on the Ontology of Design' Draft paper at: http://sawyer seminar.ucdavis.edu/files/2012/12/ESCOBAR_Notes-on-the-Ontology-of-Design-Parts-I-II-_-III.pdf (last accessed on the 22nd of January 2015), p. 66.

37   P. Descola, *Beyond Nature and Culture*, 2013, Chicago, IL: The University of Chicago Press.

38   For a history of the development of the law and geography project, see L. Bennett and A. Layard, 'Geographies: Becoming Spatial Detectives' 2015, *Geography Compass*, 1–16.

39   J. Holder and C. Harrison, 'Connecting Law and Geography' in J. Holder and C. Harrison (eds), *Law and Geography*, 2003, Oxford: Oxford University Press, 3–16, p. 5.

40   N. Blomley, 'Simplification is complicated: property, nature, and the rivers of law', *Environment and Planning A*, 2008, vol. 40, 1825–42, p. 1838.

41   Ibid.

42   N. Blomley, 'Making Private Property: Enclosure, Common Right and the Work of Hedges', *Rural History*, 2007, pp. 1–21.

43   Ibid., p. 16.

44   N. Blomley, 'From 'What?' to 'So What?' Law and Geography in Retrospect' in J. Holder and C. Harrison, *Law and Geography*, 2003, Oxford: Oxford University Press, pp. 17–34.

45   Ibid., p. 32.

46   Ibid., p.31.

47   D. Delaney, 'Tracing displacements: or evictions in the nomosphere', *Environment and Planning D*, 2004, vol. 22, 847–60 and D. Delaney, *The Spatial,*

the Legal and the Pragmatics of World-making: nomospheric investigations, 2010, Routledge: New York.

48   D. Delaney, *The Spatial, the Legal and the Pragmatics of World-making: nomospheric investigations*, 2010, Routledge: New York, p. 25.

49   D. Delaney, *Law and Nature*, 2003, Cambridge: Cambridge University Press, pp. 403–4.

50   F. Von Benda-Beckmann, K. Von Benda-Beckmann and A. Griffths, 'Space and Legal Pluralism: an Introduction' in F. Von Benda-Beckmann, K. Von Benda-Beckmann and A. Griffths (eds), *Spatializing Law: An anthropological Geography of Law and Society*, 2009, Farnham: Ashgate, 1–30, p. 4.

51   M. G. Wiber, 'The spatial and temporal role of law in natural resource management: the impact of state regulation of fishing space' in F. Von Benda-Beckmann, K. Von Benda-Beckmann and A. Griffths (eds), *Spatializing Law: An anthropological Geography of Law and Society*, 2009, Farnham: Ashgate, 75–94.

52   M. Harm Benson, 'Mining sacred space: law's enactment of competing ontologies in the American West', *Environment and Planning A*, 2012, vol.44, 1443–58.

53   Ibid.

54   I. Braverman, 'Uprooting Identities: The Regulation of Olives in the Occupied West Bank', *POLAR: The Political and Legal Anthropology Review*, 2009, vol. 32, 237–63.

55   M. Pieraccini, 'Property Pluralism and the partial reflexivity of conservation law: the case of upland commons in England and Wales', *Journal of Human Rights and the Environment*, vol. 3, pp. 273–87.

56   See for instance the work by Strang 2011 about claims to ownership of water in Australia. V. Strang, 'Fluid Forms: Owning Water in Australia' in V. Strang and M. Busse (eds), *Ownership and Appropriation*, 2011, Oxford: Berg, pp. 171–95.

57   Interestingly, if for Palestinian the olive tree embodies their connection to the land, for Zionists the steadfastness of the tree is an indicator of the fact that it requires unsophisticated cultivation techniques. The olive tree is used in Zionist narratives to symbolise Palestinian agricultural backwardness and their inability to cultivate something requiring more care.

58   Ibid., p. 31.

59   This is so unless management agreements or other forms of land management exist limiting their exercise.

60   Ecclesiastical Commissioners. Griffiths and Others, Cardiganshire Assizes, 1875 (Nichols and Sons, Westminster 1875). Griffiths and other commoners claimed that sheepwalks in the manor of Llandewibrefi were the graziers' own freehold mountain land but the court found in favour of the lord of the manor pointing to evidence of manor court presentments regulating common grazing.

61   Above n 53.

62   I. Braverman, N. Blomley, D. Delanye and A. Kedar, 'Introduction: Expanding the Spaces of Law' in I. Braverman, N. Blomley, D. Delanye and A. Kedar, *The Expanding Spaces of Law*, Stanford, CA: Stanford University Press, 1–29.

Index